Electrics

Marshall Cavendish

CONTENTS

Editor Fred Milson
Designer Chris Walker
Production Controller Steve Roberts
Typesetting ABM Typographics Limited, Hull

Produced by Marshall Cavendish Books Limited
58 Old Compton Street
London W1V 5PA

ISBN 0 86307 353 0

Printed and bound in Milan, Italy by New Interlitho

INTRODUCTION

A clear understanding of how your car's electrical system works is essential for the home mechanic, whether you want to track down faults or add on extras such as fog or spotlights. In simple, everyday language, *Car Electrics* explains how the system operates, tells you how to wire up extra driving lights and shows you what to do when important electrical components, such as indicators and brake lights fail to work

Tracking down electrical faults

Tracking down almost any electrical fault simply requires a little basic knowledge and a methodical checking procedure — here demonstrated on a faulty number plate lamp

Electrical equipment may fail to work for several reasons. The component itself may be faulty or worn, a wiring connection may be loose or dirty, the metal around the earth wire might be rusty, causing a bad earth, the core of the wire could be broken from being bent too often, or the plastic insulation of the wire could be damaged or worn. This would allow the current to escape to earth and so create a short circuit.

The type of fault can sometimes give you an idea as to its cause. For instance, dim lights are often caused by a bad earth, a blown fuse is often the sign of a short circuit and a light or other component that flickers or works only intermittently suggests a loose connection.

But whatever the problem, a few quick checks will often help you to find it — without

When to do this job
If an electrical system fails or works only intermittently
If a fuse blows

What this job involves
Using test equipment
Systematic testing of an electric circuit

Points to watch
If you run into trouble get a circuit diagram for your car and trace out the circuit you are testing
Test lamps only work as well as their earth connection allows. Make sure that your crocodile clip gets a good grip on clean metal
If a wire is earthed to the car body clean the metal with emery paper and coat the connection with grease

To do this job
Tools: Test lamp; circuit tester; length of wire
Materials: New fuses (maybe); insulation tape (maybe)
Degree of difficulty: Needs patient and methodical approach

If you have the job professionally done . . .
Does electrical component now work properly?
Have fuses stopped blowing?

even getting out your tool box.

For instance, if one sidelight is not working but the other three are, the switch must be working and the fault is likely to be in the bulb or its holder or the wiring to that bulb. Or if you notice that, for example, both the horn and the interior light have stopped working, check whether they share the same fuse. If they do, a blown or loose fuse is likely to be to blame.

If you have recently had the boot heavily laden and then notice that a rear light has stopped working, see whether you have accidentally pulled off a wire in the boot.

When only one brake light bulb works, it cannot be the switch that is faulty. But if neither brake light works, the switch could be the culprit — or, just possibly, both bulbs have failed at the same time. If the same fuse keeps on blowing, see which components work through that fuse and turn them on one at a time. Then you can tell which one is making the fuse blow. When you know which component is at fault, you can check that component and its wiring to find the fault. If the fuse only blows when all the components are turned on, check the fuse rating.

FACT FILE

Car wiring basics

An electrical device will work only if a complete electric circuit is made. Batteries have two terminals, one positive (+) and one negative (—). To make a circuit a wire must run from one terminal to the electrical device; this is the *live* wire. Another wire must take the current back to the other battery terminal; in cars, this is called the *earth* wire. Some materials, usually metals like copper and steel, allow electricity to flow through them and are called *conductors*. Other materials like plastics and glass will not let electricity pass and are called *insulators*.

To save wiring in a car, the steel car body itself is used as one of the conductors in the circuit. One battery terminal has a braided metal strap or cable which is bolted to the body nearby. In almost every modern car, this is the negative (—) terminal. The car body is thus used as the negative conductor, and the electrical system is described as *negative-earthed*.

In this system, many electrical devices have short wires (usually black) which are attached to the car's body. These are the earth wires which pass the current to the body and thus back to the earthed terminal of the battery. Other electrical components are bolted directly to the car body, so do not need a separate wire connecting them to the body.

In a few older cars, it is the positive (+) terminal of the battery which is connected by a strap to the car body. This system is described as *positive-earthed*.

In the tests described here, it makes no difference whether your car is negative-earthed or positive-earthed. This only becomes important when fitting radios or alternators.

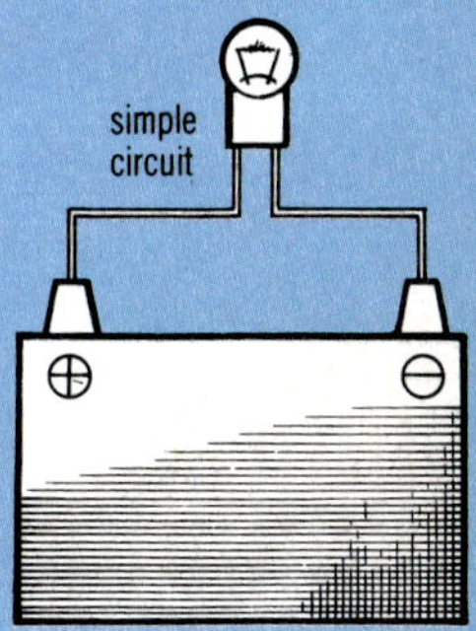

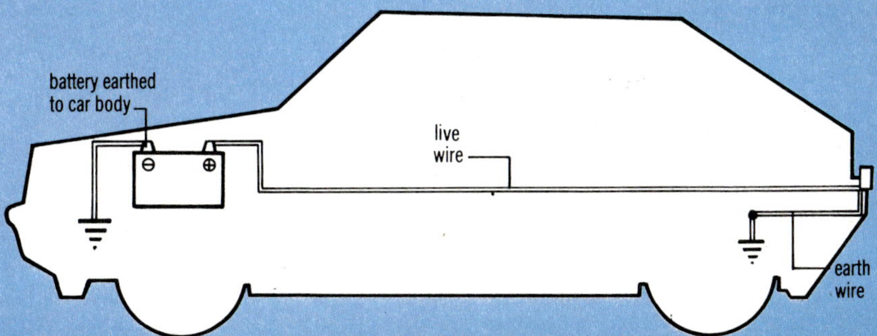

A light circuit (top) in theory and (below) in practice

FACT FILE

Switches and fuses

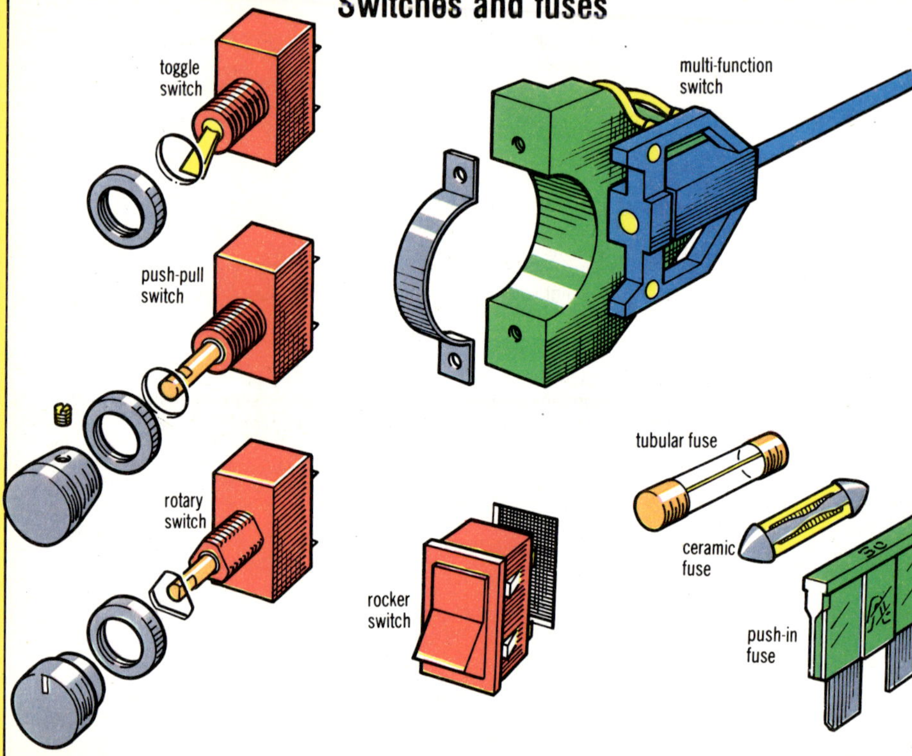

Switches
A switch is a device which breaks a circuit to prevent electricity from flowing. Not all switches are on the instrument panel for the driver to operate. An electric radiator fan, for instance, has a thermostatic switch in the cooling system which turns the fan on when the engine gets too hot. The brake lights are operated either by a mechanical switch underneath the brake pedal or by a hydraulic switch somewhere in the brake line.

There are various kinds of switches which can be operated by hand. They include toggle switches, rocker switches, rotary switches, and multi-function switches which control more than one piece of equipment, such as the lights and the horn.

Fuses
A fuse is deliberately designed to be the weakest link in a circuit. It often consists of a glass tube with a metal contact at each end containing a very thin wire which runs from one contact to the other. Fuses are designed to burn out or blow when they are asked to carry too much current. So, if a fault somewhere overloads the circuit, what you get is a blown fuse — not a car fire. Different ratings of fuses are used, depending on the number and type of electrical components in the circuit.

When a fuse blows, it usually means that there is a fault in the circuit which should be mended before a new fuse is fitted. Often, for example, blown fuses are caused by a live wire or a live connector touching the car's body and making a short circuit. (You can put this right by using insulation tape to cover the bare wire.) If no fault is found, a new fuse can be fitted because fuses do sometimes wear out. But if a newly fitted fuse blows there is definitely something wrong.

Your driver's handbook will show you where to find the main fuseholder, which fuse serves which components and what ratings the fuses are. Some components may have their own fuses fitted somewhere along their supply wires and not in the fuseholder; your manual will tell you this, too.

If you renew a fuse, always use one of the same rating as the original. A higher-rated fuse will not cure a circuit fault — in fact, it might worsen it.

STEP 1 MAKE A VISUAL CHECK

Some cars have two number plate lights, others have one. In a two bulb system, if one bulb is lit but not the other, the fault is probably a blown bulb. For both systems, look also to see if all four sidelights, as well as the number plate lights, have stopped working. If this is the case then the fault is probably in the sidelight switch or the fuse — all these lights are almost always served by the same switch and fuse circuit.

1. First check for a blown bulb

FACT FILE

Cheap test equipment

You can make a test lamp for use on a car quite easily from a 12-volt light bulb. You will also need a bulbholder (available from an accessory shop or car breaker) and a 3 ft or 1 m length of insulated electrical wire.

Cut the wire in half, bare the ends of insulation, and connect the two pieces to the terminals of the bulbholder. Some bulbholders have screw terminals; if not, solder the connections. On the end of one wire, screw or solder a crocodile clip. On the other wire solder a sharp point, such as a heavy pin or dart.

If the clip is connected to a live terminal while the point is touched against bare metal on any part of the car body, the lamp will light.

Alternatively, you can buy a test lamp quite inexpensively at an accessory shop. This can look like a screwdriver with a clear handle containing a light bulb. At the end of the 'screwdriver' there is a probe which can be used to scratch through dirt or rust to make a good electrical contact. It can also be pushed through the insulation around a wire to touch the wire inside. Attached to the handle is a wire with a crocodile clip at the end.

A circuit tester (or continuity tester) is similar to a test lamp, but has its own battery. This allows it to be used to test items which have been removed from the car, or to test for breaks in a circuit without having to leave the equipment switched on.

A circuit tester also has a sharp testing probe and a wire with a crocodile clip, but its body is fatter than a screwdriver's to make room for a battery.

Test lamps are sometimes referred to as circuit testers but, for clarity, this handbook uses 'circuit tester' only to refer to the type with its own battery.

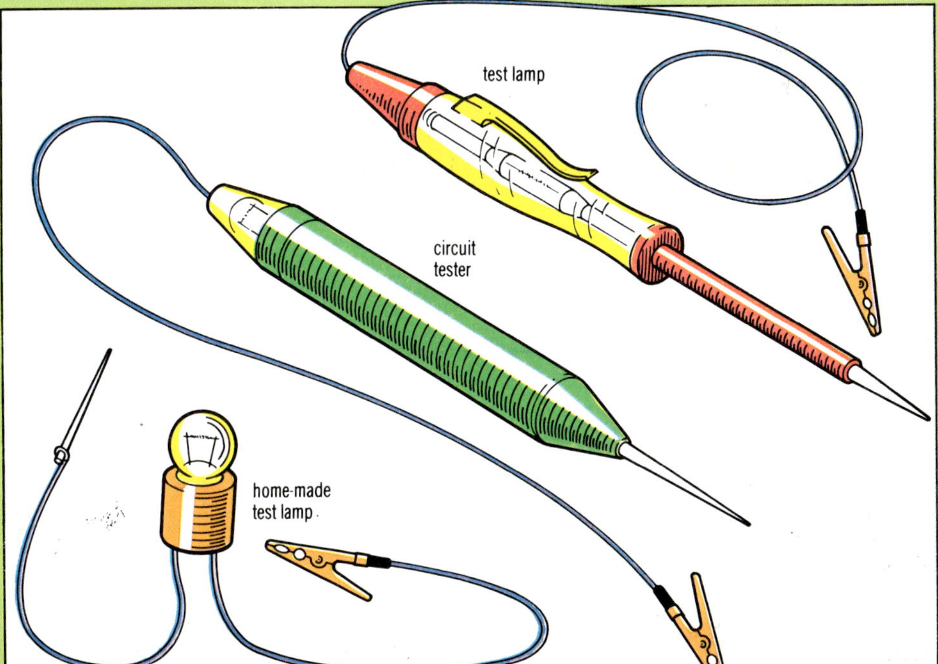

STEP 2 CHECK THE FUSE

The fuseholder is a small box, usually with an easily detachable cover, with the fuses mounted side by side inside (**fig 1**). It is usually fairly close to the battery, on the engine side of the bulkhead or the passenger side. Since locations vary, check in your manual if it is not immediately obvious.

Look in the manual to see which fuse serves the number plate lamp. Sometimes the fuseholder itself has this information printed on it.

If there is another fuse marked with the same rating in the fuseholder, do a quick test by changing the two fuses over. If the new fuse makes the component work again, you will know that either the old fuse needs to be replaced, or it was making a bad connection in the holder.

You can test a fuse where it is by using the test lamp. Fasten the crocodile clip to earth and turn the sidelight switch on to feed power into the circuit.

Then touch the test lamp probe, in turn, to both of the clips which hold each end of the fuse (**fig 2**). If the test lamp lights when the probe is in both these positions, the fuse works and is in good contact with the fuseholder clips. If the lamp lights only when touched on one end of the fuse, make sure the fuse is firmly held in its clip at the opposite end — the fuse may be intact but is not making electrical contact with its metal holder. Clean the holder terminal clips and the fuse itself, make sure the fuse is a good tight fit in the holder and try again.

If you are still not sure whether the fuse is working, take it out to test it. This time, use the circuit tester. Fasten the clip to one end of the fuse and touch the probe to the other end (**fig 3**). If the fuse is working, the circuit tester will light.

1. A typical fuseholder

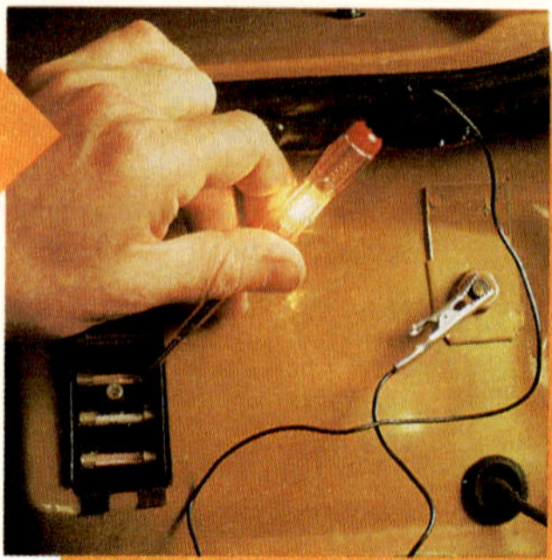

2. Checking with test lamp

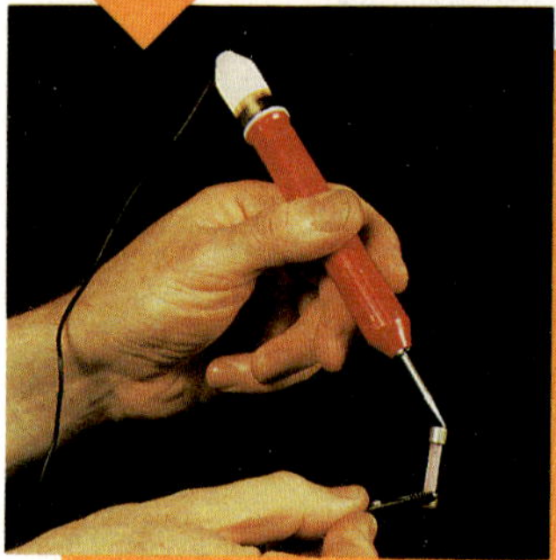

3. Using a circuit tester

STEP 3 CHECK THE BULB

Remove the bulb (see Fact File — Types of bulb) and look at it against the sky — you may be able to see whether the filament has broken. If you cannot tell, test it with a length of insulated wire. Hold one end of the wire to one battery terminal and the other end to the metal body of the bulb. Touch the base terminal of the bulb to the other battery terminal (**fig 1**). If the bulb does not light you will have to renew it.

1. Testing a bulb at the battery with a length of insulated wire

TIP

Easy way out

Sometimes a bulb which is recessed in the bulbholder can be difficult to turn when you are trying to take it out because there is not enough bulb to get hold of. A piece of Blu-Tack or a loop of sticky tape pressed on to the bulb will help you to get a grip on the glass envelope of the bulb. If your hands are oily you will find it more difficult to get a firm hold of an awkward bulb.

FACT FILE

Using test equipment

When using a test lamp, the circuit or component you are testing must be turned **on.**

If you are testing a **live** terminal or live wire, first fix the crocodile clip to earth. You can use the earth terminal of the battery, or any part of the car body where the clip can make a firm connection with clean, bright metal.

Now touch the probe to the terminal or wire you want to test. The bulb in the test lamp should light up.

Usually you will have to test at more than one place — for example, at both ends of a suspect wire — to find the fault.

To test an **earth** connection, first fasten the crocodile clip to a live point. This can be either the live terminal of the battery, or any live connection in the car's wiring such as the live terminal of a light unit or a live point in the fuse holder.

Now touch the probe to the component you want to test. The lamp should light up.

Again, you will probably have to test at more than one point to see where the break in continuity is.

Circuit tester

When using a circuit tester, the circuit you are working on must be switched **off** — otherwise power from the car's battery will burn out the bulb of the circuit tester.

You can use the circuit tester for testing components away from the car. For example, to test a fuse you simply clip the tester to one end of the fuse and touch the probe to the other. The bulb will light up if the fuse is working.

Another use of the circuit tester is to check that there is good electrical contact — for example, between a lightbulb and its holder. Clip the circuit tester to the metal of the bulbholder and then touch the probe to the metal body of the bulb. If the bulb is making good electrical contact with the holder the circuit tester will light up.

Similarly, to check whether there is a break in an electrical wire, fix the crocodile clip to one end of the wire and touch the probe to the other. The bulb will light if the wire is unbroken.

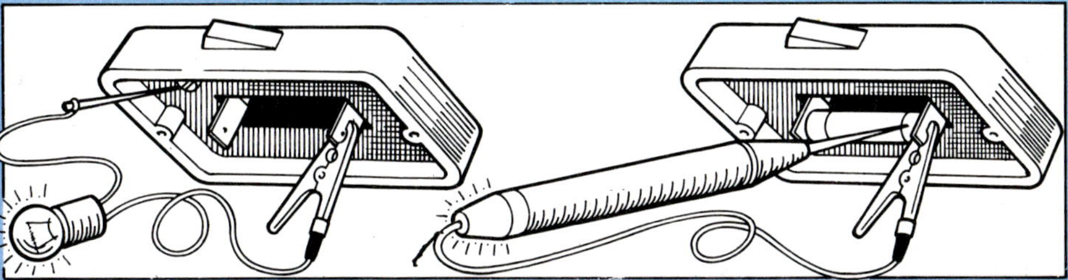

FACT FILE

Types of bulb

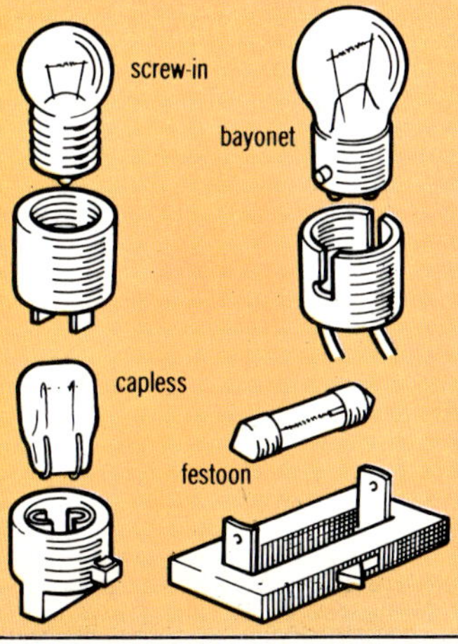

There are four types of bulb commonly used in car lights. Your car may use a different type of light fixing in the sidelights and number plate light, for instance.

Some bulbs screw into the bulbholder and can be removed by turning them anticlockwise.

The capless type of bulb is a push fit in the bulbholder. To take it out, do not try to turn the bulb but just pull it.

The bayonet type can be taken out by pushing it in slightly and then turning it anticlockwise. Spring pressure will then push the bulb outwards.

Festoon bulbs are tubular in shape with pointed ends and are held in the bulbholder by a spring clip at each end. To take out a festoon bulb, pull one spring clip gently away from that end, then ease the bulb in the other direction with your other hand.

Whatever the type, there will be two contacts or terminals. One will be live, the other earth. It is usually clear which is which. With festoon bulbs the earth is the terminal connected to the metal holder, while bayonet and screw bulbs are earthed through their metal casing.

STEP 4 CHECK THE CONNECTIONS

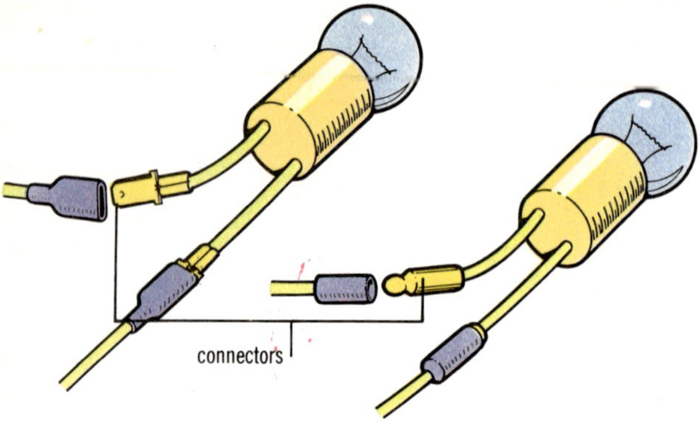

1. Often connectors look tight but are not making good contact

The light may have stopped working because the wires to it have worked loose. To test for this, turn on the sidelight switch. Check that all the connectors between wires and bulbholder are tight. (See **fig 1** for various types you might encounter.) Do this by pulling the connectors apart then pushing them firmly back again. While disconnected make sure the terminals are clean and free from corrosion — rub them lightly with emery paper. Dirty connectors may fit tightly together but may not make electrical contact. If poor contact was the problem the number plate light will now come on.

STEP 5 CHECK THE BULB IS FIRMLY HELD

A bulb may seem to be tight in the bulbholder but, like wiring connections, electrical contact may not be made. To check this, leave the sidelight switch on. Try wobbling the bulb with a fingertip to see if it lights up.

If it does, the bulbholder is faulty. Cleaning off any corrosion may fix it; otherwise it will have to be renewed.

If the bulb does not light up, the fault may still be in the bulbholder. To make sure, you will need to check both whether power is reaching it (see Step 7) and whether it is properly earthed (see Step 6). If these connections are both in order, then the bulbholder is at fault.

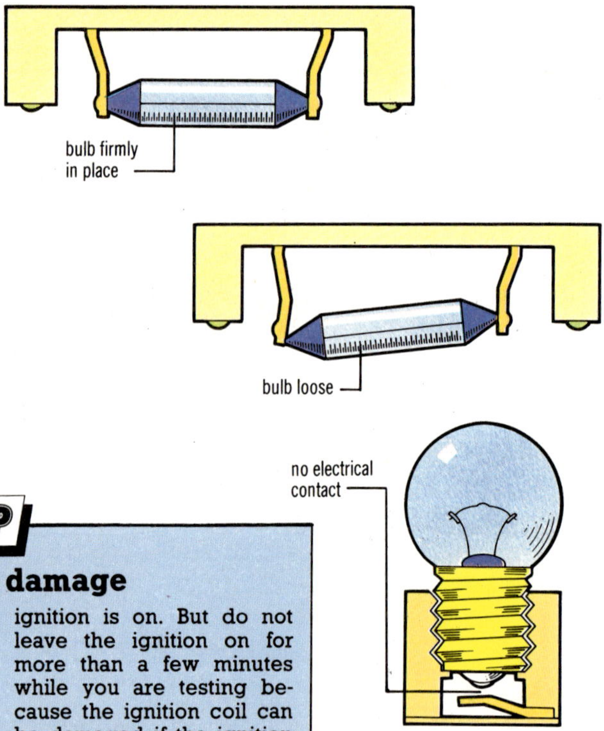

1. Bulbs which are not tight in their holders may not work at all or may flicker

TIP

Avoiding damage

Some electrical components — windscreen wipers, for example — work only when the ignition is switched on. If you are testing one of these components (or its switch, or the fuse in the circuit concerned) make sure that the ignition is on. But do not leave the ignition on for more than a few minutes while you are testing because the ignition coil can be damaged if the ignition switch is on the 'on' position for too long without the engine running.

STEP 6 CHECK THE EARTH

With the number plate bulb removed and the power off, clip one end of the circuit tester to a good earth. Then touch the probe of the tester to the inside of the bulbholder, which grips the bulb and earths it (**fig 1**). (In the case of a festoon bulb, touch the metal body of the bulbholder.) If the tester lights, the bulbholder is properly earthed. If the tester does not light, the earthing of the bulbholder is faulty.

Make sure you have cleaned all rust and dirt off the earth terminal of the bulbholder.

As **fig 2** shows, you may be able to fasten one end of a length of wire to the earth terminal of the bulbholder and the other end to an earth point, such as a nearby bolt, on the body of the car. Then test the bulbholder again to see if it is now earthed. If the bulbholder is properly earthed but the number plate lamp still fails to light, the fault is probably in the power supply wire.

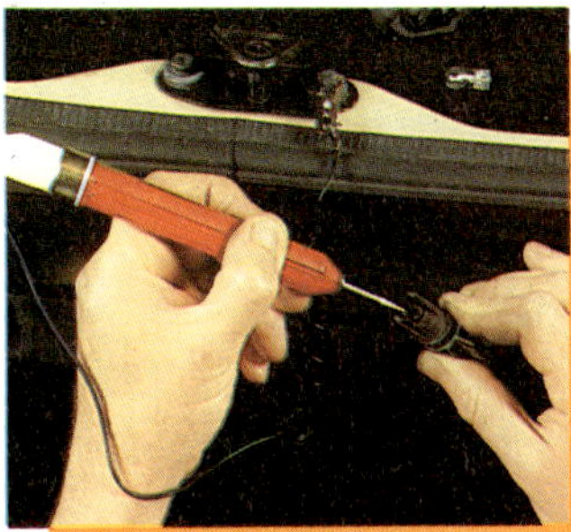

1. Testing earth terminal

2. Fixing extra earth wire

STEP 7 CHECK THE POWER SUPPLY

Turn on the sidelight switch. Clip the test lamp to an earth point. Touch the probe to the live terminal of the number plate light bulbholder (**fig 1**). If the test lamp lights, current is reaching the bulbholder and you should check again that the bulb is firmly held and that the wiring connections are tight, since there is no other reason why it should not work.

If the test lamp does not light, current is not reaching the bulbholder. So you will need to check the power feed wire (see Step 9).

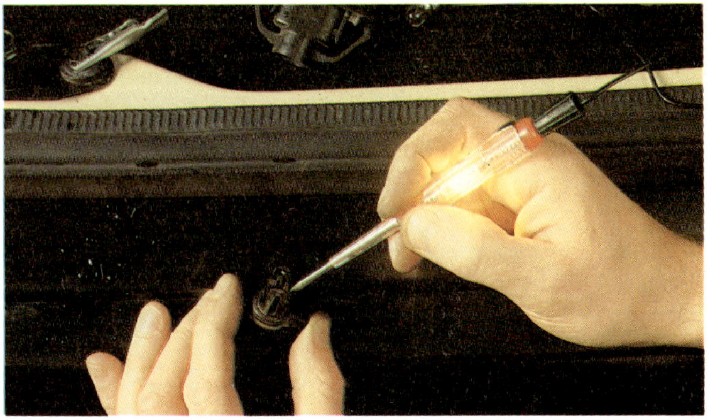

1. Using the test lamp to check the power supply wire

STEP 8 CHECK THE SWITCH

If you can reach the back of the switch which operates the number plate light and the sidelights, use the test lamp to check whether there is current at both terminals of the switch when the switch is turned on (**fig 1**). If one terminal is live but not the other, the switch is faulty and must be renewed. If neither terminal is live, current is not reaching the switch and you should check the supply wire that comes from the fuseholder.

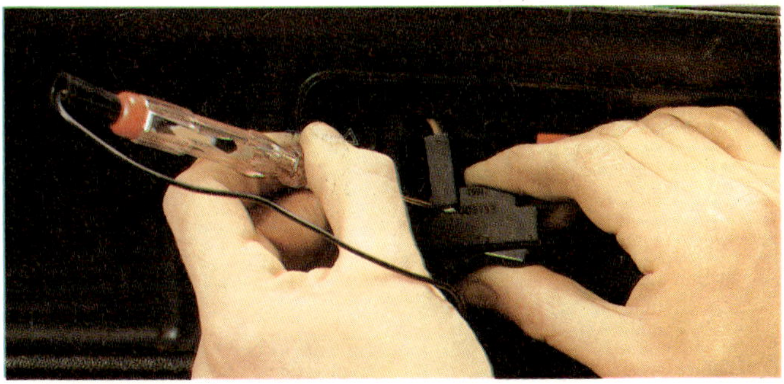

1. Testing the terminals at the back of the switch

Circuit tester

A circuit tester has its own power supply (a 3 V battery) so it can only be used to test the *continuity* of a circuit — that is, whether it is continuous, with no breaks at any point.

Use it to test the continuity of small parts of a circuit, or where there is no power from the car battery.

Examples

Is the component working?

Virtually every component in the car, including motors, will give a continuous reading when tested on a circuit tester as long as it is working.

Disconnect the component from its supply and test across its terminals (**fig 2**).

If the tester lights up it simply proves that there is no internal break in the circuit — it does not prove that the component will work, except for simple devices such as fuses and bulbs. But if you get no reading between the terminals, the component will not work.

This does not apply to HT leads, whose high resistance cuts down the low voltage of the tester to almost nothing.

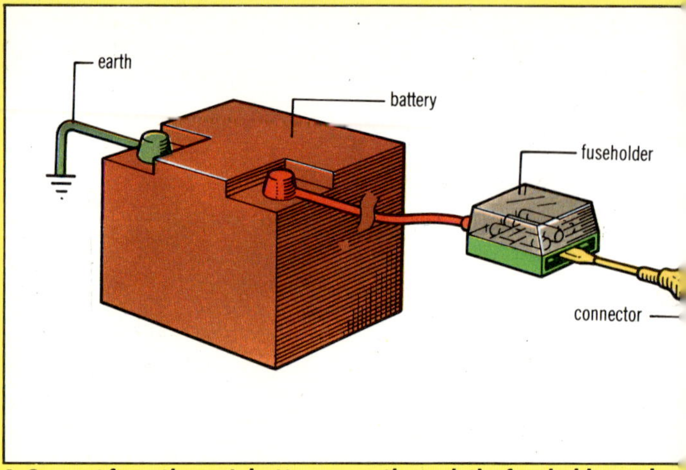

1. Current from the car's battery goes through the fuseholder and

Which wire goes where?

You are often faced with wires which disappear into a loom, and are difficult to track down. But if you can get at both ends of a group of wires (disconnect them first), you can use the circuit tester to prove which is connected to which (**fig 3**).

The lamp lights when both ends are connected to a continuous wire.

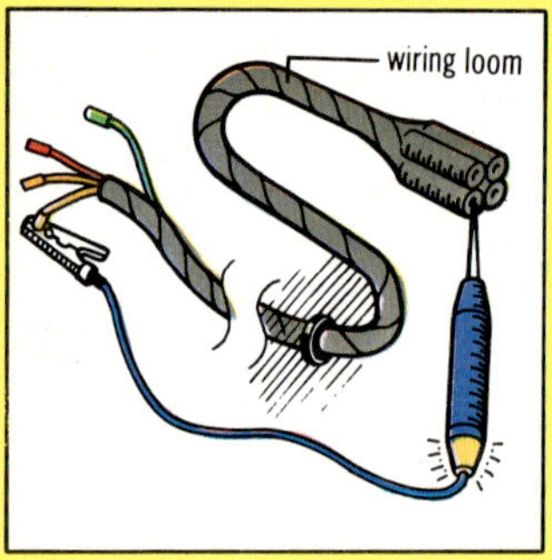

3. Testing a wire for breaks

2. Checking the continuity between the terminals of a motor

Test lamp

A test lamp must be of the same voltage as the car battery, which it uses for its power. Its purpose is to check whether or not current is reaching various parts of the circuit.

Examples

Is live power reaching the fuse?

Connect the lamp between the connector and an earth point (**fig 4**). This completes the circuit, so if power is getting through the lamp should light. If not, either your earth point is bad or the power is not getting through. Check that the earth is good by testing a known live point, such as the live terminal of the battery.

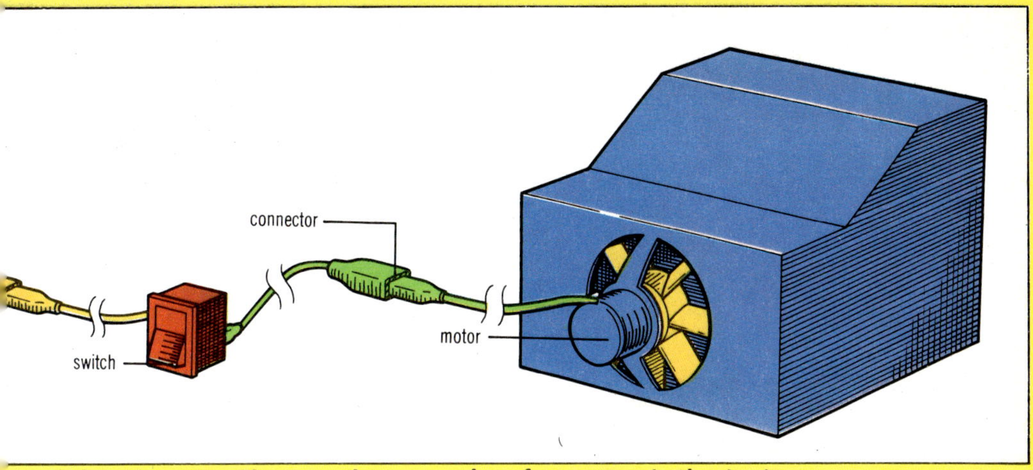

switch to the motor. There may be any number of connectors in the circuit

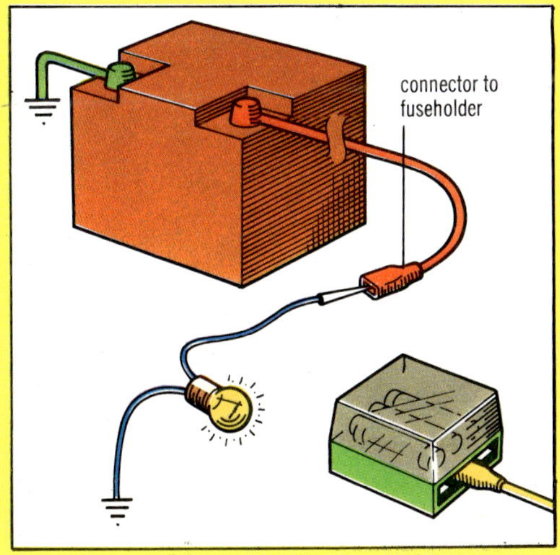

4. Testing a connector to the fuseholder

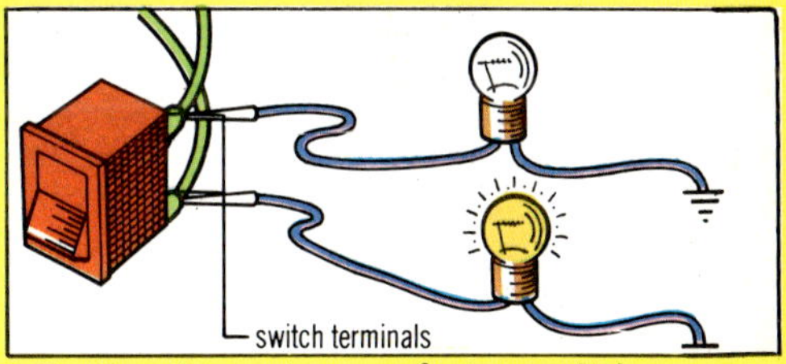

5. Testing the power supply to a switch

6. Testing whether the switch works

You can continue through the circuit stage by stage, checking at each point whether the power is reaching it. Wherever the lamp lights there is power. Remember to keep the switch on if you test beyond it.

Is the switch working?
To tell which of the two wires to the switch is live, test both with the switch off (**fig 5**).

The one which lights up the lamp must be the live feed. If neither lights, power cannot be reaching the switch.

Next, connect only the live feed to the switch, switch on and test at the other terminal (**fig 6**).

If the lamp does not light, the switch must be faulty. If it does, the switch works. Prove the point by switching the lamp on and off using the switch itself but reconnect it before continuing.

Where is the break?
If power is getting through the switch but not reaching the connector (**fig 7**), the wire between the two must be broken somewhere.

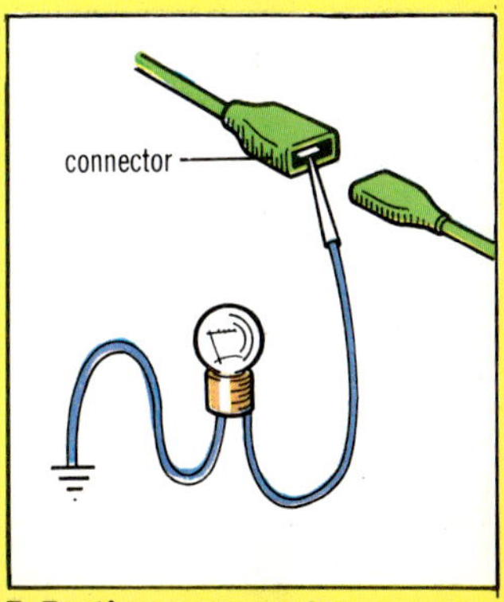

7. Testing a connector

STEP 9 CHECK THE POWER FEED WIRE

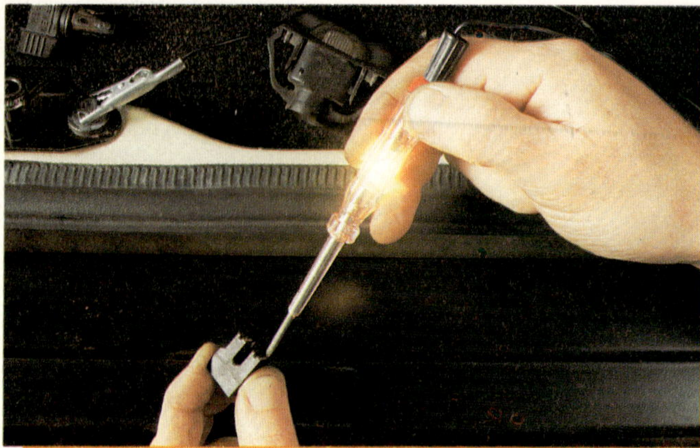

1. Testing that current is reaching the connector

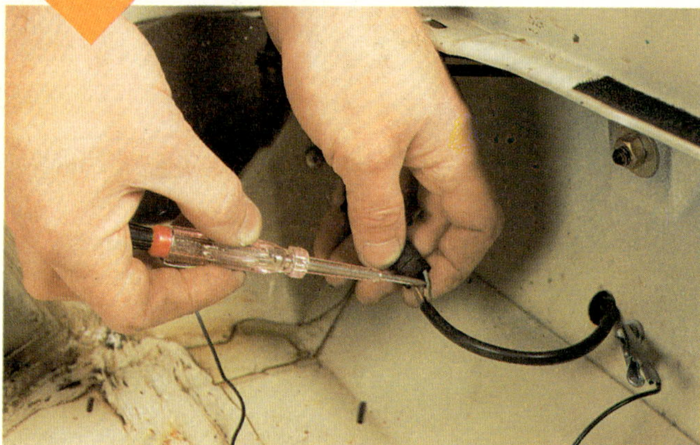

2. Probing through the insulation to test the wire inside

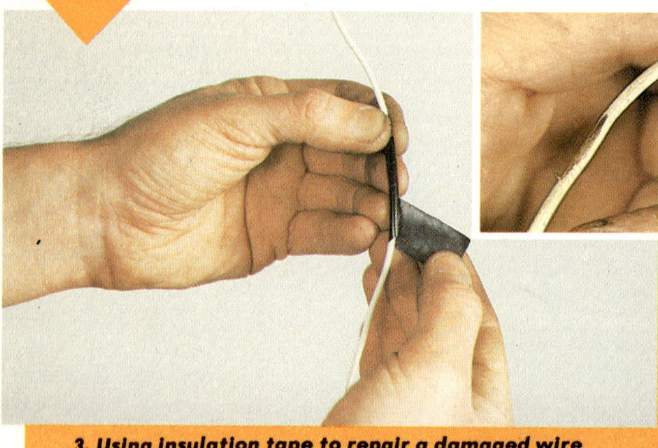

3. Using insulation tape to repair a damaged wire

When you know that the switch and fuse are working and live, the fault must be a break or short circuit somewhere along the power feed wire leading to the number plate lamp.

Make sure that any connectors in the wire are tight. Turn on the sidelight switch and clip the test lamp to earth. Then use the probe to test along the wire until you find a point in the length of wire where there is current and the test lamp lights.

The car's wiring is colour-coded, so this will help you to trace wires. Often a wire disappears through the bulkhead or into a large cluster of wires which are bundled together. Generally, a wire stays the same colour along its whole length, so you should be able to identify the wire even though you may have lost sight of it at some stage.

Test any connectors in the line either by peeling back the insulating cover and touching the probe to the metal inside (**fig 1**), or by pulling the connector apart and touching the probe to the connector terminal on the end of the wire which feeds the number plate light. Test the wire itself by clipping the test lamp to earth and pushing the probe through the insulation of the wire to touch the wire inside (**fig 2**).

Remember never to let any live connector touch the bare metal of the car — the short circuit could burn you.

When you find that there is current in one part of the wire but not at the next point along the wire, the fault must lie between those two points.

Gradually track down the exact location, then examine the wire. If its insulation has worn, causing a short circuit, you can patch it with insulating tape (**fig 3**). If it has broken, you will have to cut out the damaged length and renew it — ideally, with another length of the same colour.

Fixing faulty headlights

You should repair a faulty headlight as soon as possible. Fortunately most faults are quite simple and are easy to put right

When a headlight stops working it is more than just inconvenient — driving with a headlight out of action is illegal, even in daylight. Putting in a new bulb or sealed beam should take only a few minutes, and the whole job can be easily finished within an hour.

Since bulbs (especially quartz-halogen ones) and sealed beam units are costly, it is worth checking exactly where the fault lies before you rush out and buy a new one.

There are usually clues to the problem — if only one beam on one side has failed (usually the dipped beam, which gets most use) then it is likely that a filament has burned out. If both beams fail then the earth connection is probably faulty, while both headlights failing together points to a power supply fault. Step 4 tells you how to check all these faults in detail.

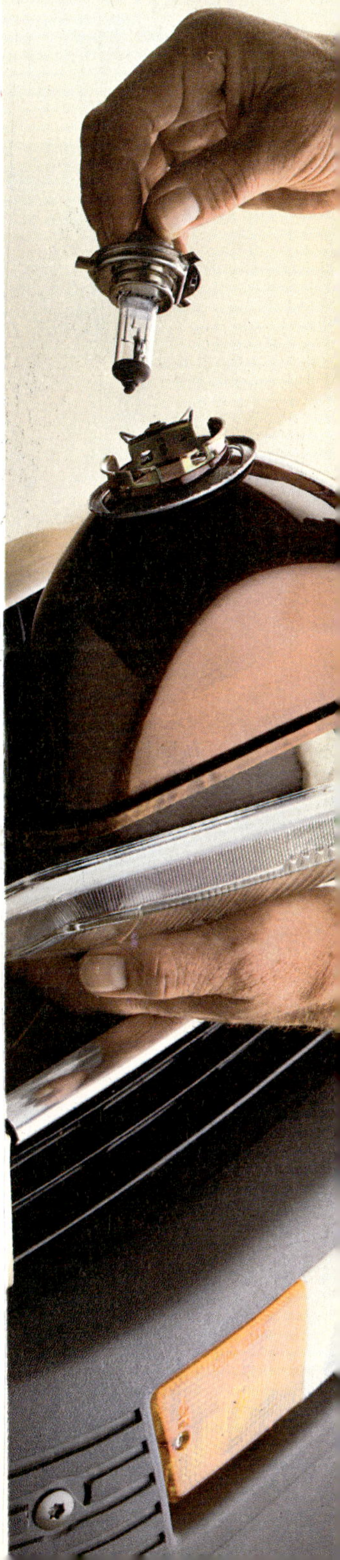

When to do this job
When a headlight bulb fails

What this job involves
Removing the headlight (if necessary)
Removing the front grille (if necessary)
Changing a headlight bulb
Changing a sealed beam headlight
Checking the unit or wiring

Points to watch
Buy new bulbs at an accessory shop, not a main dealer where they will cost twice as much

To do this job
Tools: Possibly: screwdrivers; spanners; test lamp or circuit tester
Materials: Headlight bulb or sealed beam unit
Time: Quickest about five minutes. Longest not more than one hour
Degree of difficulty: One of the easier fix it jobs

If you have the job professionally done...
Does the headlight work? Is it bright and aligned correctly so that it does not dazzle?

STEP 1 CHECK FOR ACCESS TO LIGHT

The first check is to open the bonnet and see whether you can see the back of the headlight with the wires protruding (**fig 1**). If so, the chances are that you can remove the bulb without disturbing the light itself. Check whether there is room to get the bulb out without it hitting the battery, radiator tank, etc. If anything is in the way shift it if possible and go to Step 3, otherwise you have to take the headlight out from the front (Step 2).

If you cannot see the rear of the light or your car has a sealed beam, then the headlight unit has to be taken out from the front. Look for any securing screws around the headlight bezel rim or the plastic trim (**fig 2**). These unscrew to give access to the main light fixing screws.

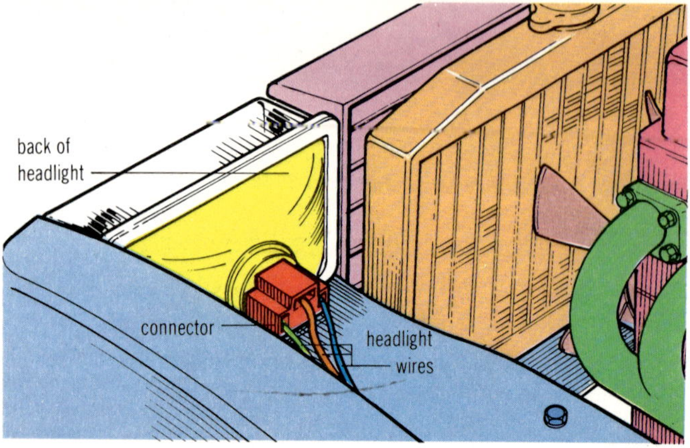

1. This type of headlight allows access from rear

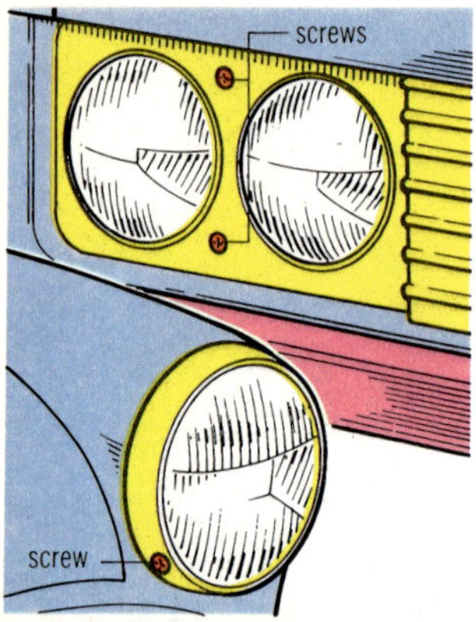

2. Trim fixing screws

If you cannot see any screws around the headlights or if the lights are set into the front grille then you have to take the grille off to get at the fixing screws. Find all the screws which hold the grille to the car and remove them one at a time. Some are held at top, bottom and sides, while others have screws at the top and lugs at the bottom which fit into slots in the front panel. Stick each screw through a sheet of cardboard to make sure you do not lose any of them. Now carefully pull the grille off, lifting it slightly if you have to, and put it somewhere where it will not get damaged. If it will not move, check for any screws hidden in recessed holes.

Watch out for any small rubber rings which fit over the lugs if your grille has them. Put them safely away. Once you have found the headlight securing screws, go on to Step 2 to remove the light unit or sealed beam.

3. Removing a grille fixing screw

TIP

Headlight conversion

If your car has a rectangular sealed beam headlight (fitted to early Allegros and Avengers for example) you will find that these are no longer available.

The cheapest way to get round the problem is to find a bulb type headlight and its electrical connectors from a similar car in the scrapyard — failing that you will have to buy a new headlight and bulb.

HOW IT WORKS

The headlight

Headlights have two beams — a main beam to illuminate the road far ahead, and a dip beam to light up the road directly in front of the car without dazzling oncoming drivers. The headlight is shaped like a bowl, the rear part having a mirrored surface to reflect light, and the front a glass lens to focus the beam.

The light for each beam is provided by a filament inside the headlight. The light reflects off the rear of the headlight and is focused by the headlight lens to form a beam of light which is concentrated on the road itself.

The main beam filament is fitted at the focus point or centre of the reflector so that the light shines straight ahead. The dip beam filament is usually offset higher up in the bulb than main, so that the light bounces off the top half of the reflector and shines down on to the road.

There are two types of headlight — the conventional bulb type, where the bulb sits in the rear of the unit and is replaceable, and the sealed beam, which is a one-piece unit cast entirely in glass and is replaced as a whole. You can tell if you have this type by the words 'sealed beam' in raised letters in the glass.

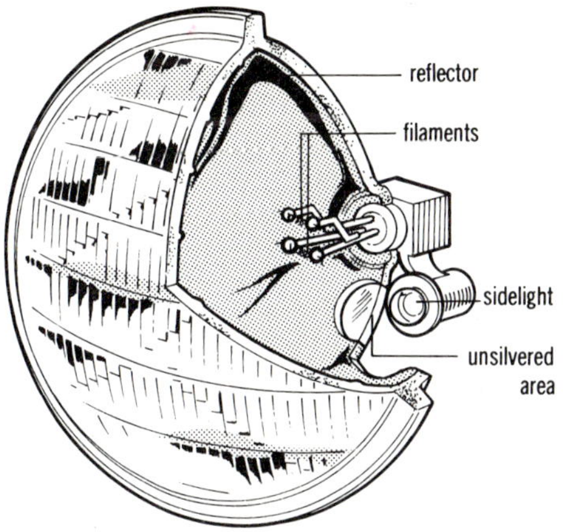

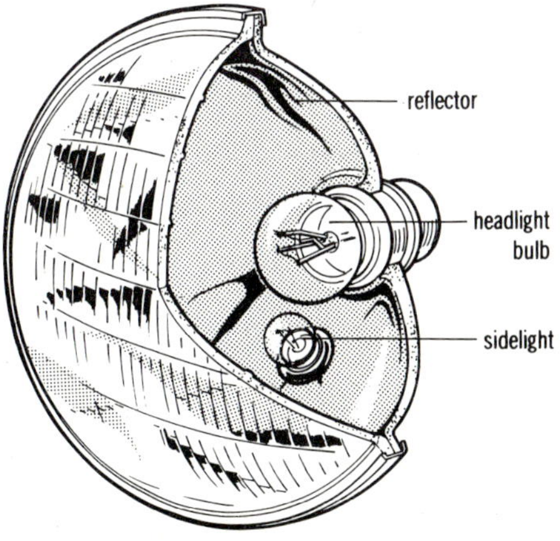

The bulb type light is fitted with one of two types of bulb — the tungsten bulb or the quartz-halogen bulb.

The tungsten bulb is made up of a glass envelope, with two tungsten filaments in it, which is filled with an inert gas such as argon. When electricity is passed through the filaments they glow, giving off light. Since there is no oxygen in the envelope they do not burn up, though the tungsten does evaporate on to the inside of the glass, blackening it.

In quartz-halogen bulbs the envelope is made of quartz instead of glass and is filled with halogen gas. This combination prevents the envelope from blackening and gives a much whiter, stronger light while drawing only slightly more power from the battery. At one time quartz-halogen bulbs were fitted only to larger, more expensive cars, but an increasing number of small cars now have them fitted as standard.

When handling halogen bulbs try to avoid touching the quartz envelope. The oils in your skin may attack the quartz and leave a deposit on the envelope so that when the bulb heats up this will form a hot spot leading to bulb failure.

Headlight bulbs are cheap and easy to renew but the bulbs often blacken with age, and water can get inside the headlight rusting the reflector.

With sealed beam headlights the unit itself is, in effect, a large bulb with the front part shaped like a headlight lens and the rear silvered like a normal reflector. Inside it are the gas and filaments, just like a normal bulb, and the whole unit is sealed so that water and dirt cannot get in and attack the reflector.

Where the sidelights are integral with the headlights they shine through a hole or a small unsilvered area in the reflector or the back of the sealed beam.

STEP 2 REMOVE THE HEADLIGHT UNIT

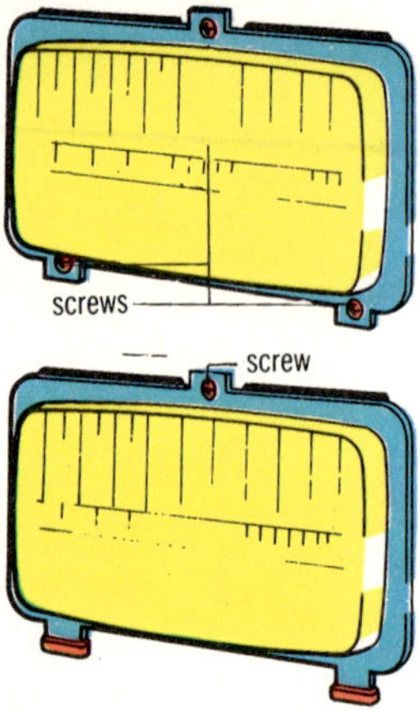

1. Headlight fixing screws

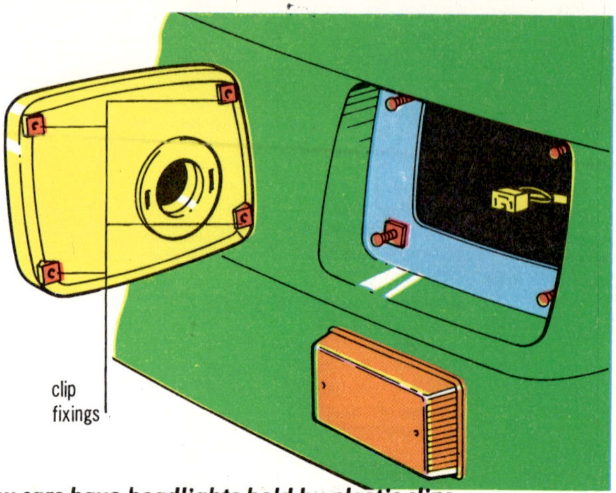

2. A few cars have headlights held by plastic clips

First disconnect the battery to avoid a short circuit.

With the grille off look closely at the headlight and look for any fixing screws, taking care not to confuse them with the spring-loaded beam adjustment screws. Rectangular lights are often held by a single screw and several lugs or by screws at the top and bottom of the headlight holding plate (**fig 1**). If there are no screws, the light is fixed by plastic lugs (**fig 2**). To remove it simply lever it forward with a thin piece of wood, and pull it out. Round units normally have three or four screws which have to be undone (**fig 3**). In some cases you will not have to take the screws out altogether. Look for keyhole slots or cut-outs around the screws — if you find them all you have to do is slacken the screws off and turn the metal ring to release the headlight unit (**fig 4**). If the screws run through a small spring you do not even have to loosen them — simply push the unit back against spring pressure and turn it.

Do not pull the unit out too far as it is still attached to the wires. Unplug the terminal block, levering it off with a screwdriver if it is tight, and release the sidelight bulbholder if it is part of the light fitting (**fig 5**).

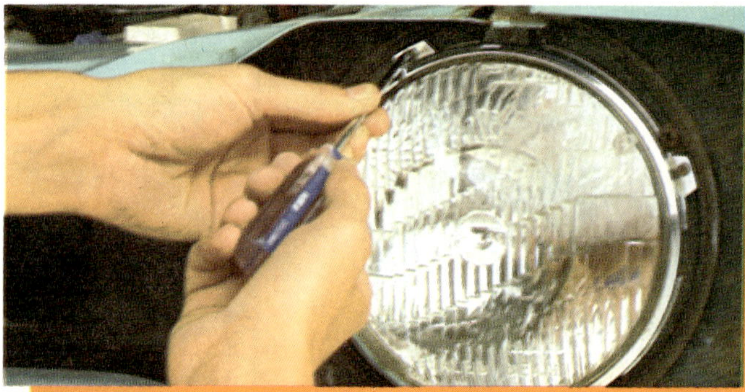

3. Undoing the headlight securing screw

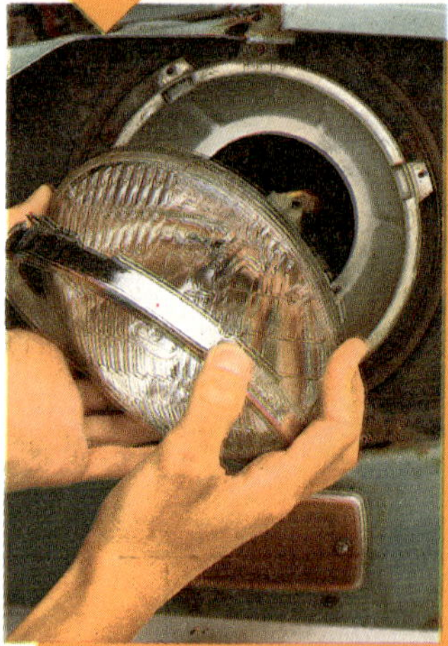

4. Lifting out the headlight

5. Pulling out sidelight

STEP 3 REMOVE THE BULB

1. Easing the connector off the headlight

2. Peeling off the protective rubber boot

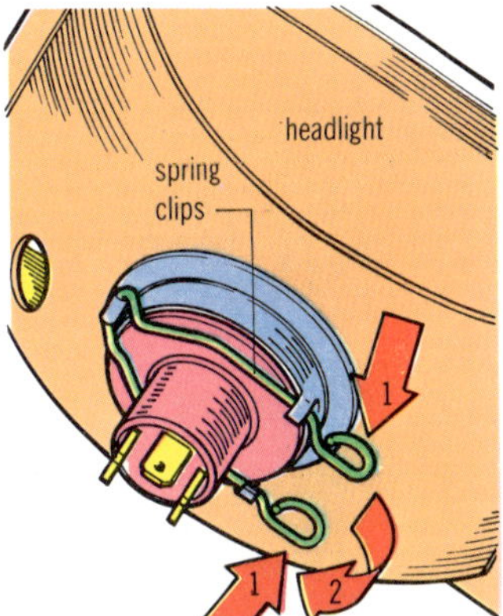

3. Spring clip bulb retainer

Once you have prised the terminal block off (**fig 1**), you can change the bulb or sealed beam.

Changing a sealed beam unit is simply a matter of putting the new one in (Step 5) but if your car uses bulbs you must remove the old one first.

First peel the rubber boot (**fig 2**) off the light — this reveals the bulb fixings. There are several ways in which the bulb can be held. There can be flat wire springs (**fig 3**) or a retaining plate which is often spring loaded (**fig 4**).

Carefully lift the springs out of their retaining grooves or, if you have a retaining plate, push it in and turn it anticlockwise. Pull the bulb free of the headlight.

While you have the bulb out look through the bulbhole and check the condition of the reflector towards the bottom of the light. Any signs of rust or lifting of the silvering (**fig 5**) mean that you should consider buying a new headlight fairly soon or the car may fail the MOT roadworthiness test. As a temporary measure, though, you can try brushing some silver paint over the damage with a small paintbrush.

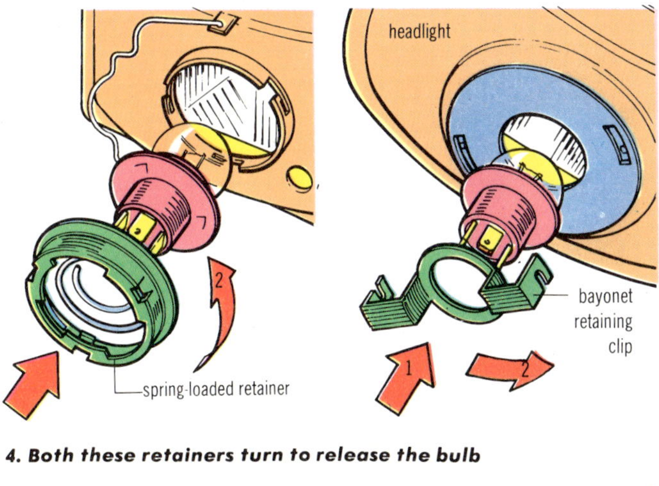

4. Both these retainers turn to release the bulb

5. Badly rusted reflector

STEP 4 CHECK THE LIGHT OR WIRING

There are several ways of checking whether or not a bulb or sealed beam has failed, and which you use depends on what equipment you have.

A burned out bulb filament is sometimes obvious because the filament is broken, but the filaments in sealed beam headlights are hard to see.

Another method is to swap the bulb or sealed beam unit for the one on the other side of the car, assuming that the other light works properly. If the apparently failed light now works, the fault is evidently in the wiring, or it may even be that one of the terminals was simply dirty, and removing it has produced a better contact. In this case just clean up the contacts with fine emery paper or WD40.

But if you have a circuit tester powered by its own battery, you can check the filaments for continuity. A bulb with a metal casing will have two flat contacts on its base. Test between the casing and each contact in turn. If the circuit tester lights on only one

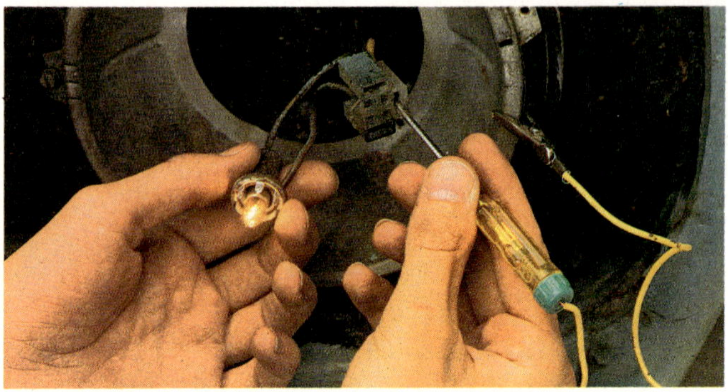

1. Testing headlight wiring — sidelight comes on as well

STEP 5 FIT NEW BULB OR SEALED BEAM UNIT

Before you fit the new bulb or sealed beam, smear a little petroleum jelly, such as Vaseline, on the bulb terminals to prevent them rusting.

If you have a halogen bulb do not touch the quartz part of it. Any finger marks will cause a hot spot which overheats when you use the light, making the bulb burn out rather quickly. If you do accidentally touch it, wipe off any marks with a rag moistened in meths and allow the bulb to dry naturally.

All double filament bulbs have some kind of pip or ear on the mounting flange to make sure the bulb is fitted the right way up in the light. These

1. Conventional tungsten (left) and quartz-halogen bulbs

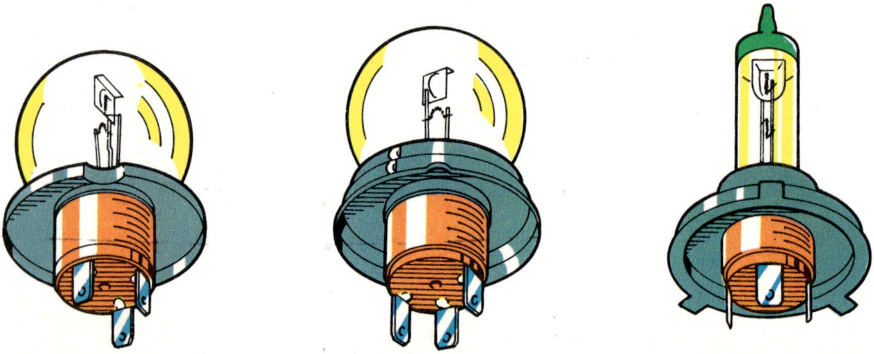

2. The three most common locating pips — make sure they align with headlight

2. Testing a halogen bulb — do not touch quartz glass

terminal then a filament has failed.

The normal bulb or sealed beam unit has three terminals — one for earth and one for each filament. You can often see which is which by looking through the glass, but if not, test between the various terminals and find out what happens. If both filaments are working the tester should light between any pair of terminals. Only one pair of terminals will make the tester light if one filament has failed. If the bulb or sealed beam is the inner one from a four headlight system, it has two terminals.

You will have to check the wiring if your tests show nothing wrong with the bulb or sealed beam unit. Use your circuit tester to check the continuity.

Headlight bulbs and sealed beam units are earthed by one of the three wires which lead to the connector. If you cannot identify which is the earth wire, you will have to check which two light a test lamp when the dip and main beam switches are on (**fig 1**). The other one will be the earth connection, and should make a circuit tester light when connected between the earth wire and a good earth point. If the earth wire is faulty in some way, check to see if you can find the place at which it is earthed and clean it up.

locating keys are shown in **fig 2**. Fit the bulb so that the ears or pip fit into the corresponding cut-out(s) in the headlamp. Give it a firm push to make sure it is properly seated, then refit the spring clips or the holding plate and check that the bulb is held firmly in position. Now refit the light unit (if you had to remove it) and the grille or headlight trim.

If your car has sealed beams then all you have to do is fit the new one in place of the old. Do not worry if it is marked 'R'. This simply means it is for a right hand drive car and can be fitted to either side of the car. Make sure that any rubber pads that were fitted are put into place in the backing plate. Check to see if they are stuck to the old unit if yours are missing. The sealed beam has castings in the back which locate into the backing plate (**fig 3**) so you cannot fit it wrongly.

Refit the headlight, remembering to replace the sidelight bulbholder if it is fitted and then fit the retaining ring and screws, but do not tighten up the screws at once. Make sure the light is seated properly in its mount, and only then do up the screws. Before you refit the grille or headlight trim, reconnect the battery and check that the lights work and are properly focussed.

Finally, refit the grille or headlight trim, making sure all the screws are tight and secure.

TIP

Lube it

Before you put the headlight or sealed beam back in it is well worth giving all the screws and headlight beam adjusters a good squirt with WD-40 or a similar water repellant spray. You should then find that if you have to take the headlight out at some time in the future, or need to adjust the light beam, the screws or adjusters will not be rusted or seized solid.

3. The sealed beam has castings to locate it

Fitting extra driving lights

Driving lights not only make your car look better, they make night driving safer by helping you see farther

Town drivers rarely need extra lights. But in the country, additional driving lights help make those dark, dangerous roads safer. To get the best from your lights it is important to set them up correctly. For maximum performance mount them above the front bumper, and make sure they are secure and free from vibration to avoid upsetting the beam adjustment.

When to do this job
When you want more or better driving lights

What this job involves
Drilling holes in bumper or body
Fitting special brackets to mount lights on
Wiring in the lights
Aligning the lights

Points to watch
Consider fitting better headlights instead of auxiliary lights because they are better aerodynamically, and they are also thief-proof
Never fit quartz-halogen bulbs in a light designed for tungsten bulbs

To do this job
Tools: Spanners or sockets; screwdrivers; pliers, electric drill (maybe); drill bits (maybe); rat-tail file (maybe); wire strippers
Materials: Wiring 28/.012 gauge; crimp-on connectors; insulators; Scotchloc connectors (maybe); rubber grommets; insulating tape or plastic sleeving; in-line fuse holder (maybe); 15 amp fuse; relay; switch (maybe); assorted nuts, bolts, washers and screws
Degree of difficulty: Drilling holes requires care, bumper removal is often hampered by rusty nuts, but wiring is not difficult with the right tools

If you have the job professionally done . . .
Do the lights work? Are the lights correctly aligned and vibration free? Do the lights go off when you dip the headlights?

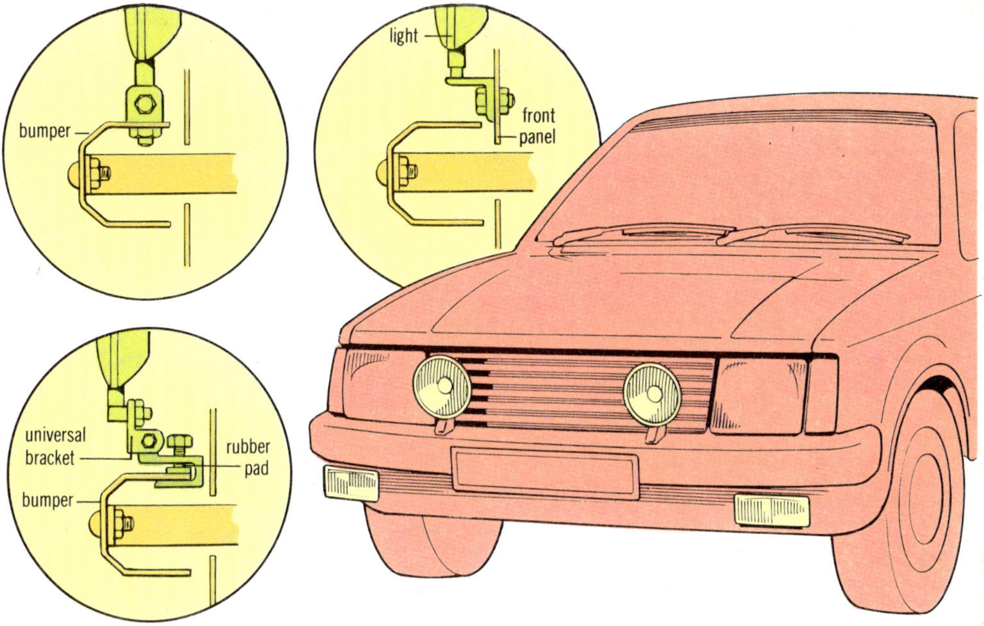

Lights can be fitted in several ways — choose whichever one best suits your car

FACT FILE

Extra lights law

Most countries have regulations covering the fitting and use of lights. Although these tend to change from time to time, any extra lights you fit must fall within the guidelines laid down by the law.

If you are in any doubt about the regulations for the type of lights you want to fit, your local authority or police force will usually help.

In the UK, for example, if you fit a pair of lights, each must be mounted at the same height and the same distance from the centreline of the car.

There are also rules covering the height at which you can fit them. If you want to use them as driving lights to supplement your headlight main beam, mount them not less than 24 in. (610 mm) and never more than 42 in. (1070 mm) from the ground. If they are set up as driving lights they must be wired so they go out when you dip the headlights.

If you fit the lights so they are permanently dipped towards the road in line with dipped beam, they can be wired independently from the headlights but must have a dashboard mounted warning light to tell you they are on. Lights wired like this can be used instead of headlights or in conjunction with them.

If you fit the lights below 24 in. (610 mm) then they can only be used in fog, falling snow or heavy rain and you need a warning light for them. Also, if your car was registered before December 1970, the distance between the inner edges of the lights must not be less than 14 in. (350 mm). Otherwise, you must fit them not more than 16 in. (400 mm) from the edge of the car, measured from the outer edge of the light.

If you fit a single light, the same fitting distances apply, and it can only be used in conjunction with the headlights — not on its own.

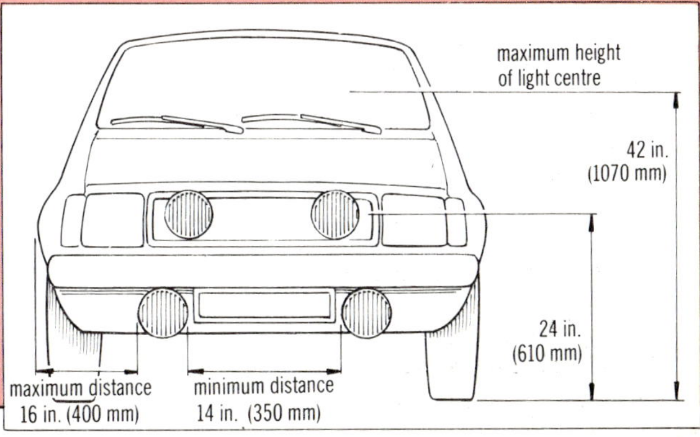

STEP 1 DECIDE WHERE TO FIT THE LIGHTS

Before you decide where to mount your new lights, you must be aware of the regulations governing extra lights (see Fact File — Extra lights law). Some of the regulations are very specific and your choice of location is restricted.

Once you know where the law allows you to mount lights, you can look at the alternatives. To some extent your choice is influenced by the type of light and mounting bracket. But whatever type you use, the mounting points must be strong and rigid to avoid the lights vibrating.

Usually, you will have to drill holes in the bumper or a front body panel — in either case you should remove the bumper first. Alternatively, if you have 'universal' clamp mounts, you can clip the brackets directly over the bumper (see Alternative step 2). This avoids removing the bumper, but it is just as easy for thieves to take off the lights as it is to fit them.

STEP 2 REMOVE THE BUMPER

On most cars, the front bumper is easily removed by undoing the fixing bolts and pulling it off.

Although mounting positions vary from car to car, most bumpers are bolted to the body in four places. The main bumper mounts are two thick steel brackets bolted to the inside of the bumper. They fit through two holes or slots in the front panel and bolt to the main frame of the car, usually under the wheel arches (**fig 1**) or inside the engine compartment. In some cases the fixing bolts go right through the inner wing, so you may need a helper to hold a spanner on one end of the bolt as you undo the nut.

Most bumpers are also fixed to the edge of the front panel or the wing. Sometimes this fixing is a bracket which is bolted to the bumper and the front panel (**fig 1**) — otherwise there are plastic clips which are screwed to the panel (**fig 1**) and the bumper wraparound section is a push-fit on to them. A third fixing type uses a long self-tapping screw which screws into the bumper from the inside of the front panel (**fig 1**).

Take all the bolts out — watch out as you do this as there are often earth connections attached to them (**fig 3**). Pull the bumper free — if it sticks check that it is not snagging anywhere, or that you have missed any bolts. Now go on to Step 3 and drill holes for the lights or brackets.

When you refit the bumper make sure that you only put the bolts in finger-tight at first. Remember also to refit any earth connections. Then shift the bumper around until it is straight and tighten up all the bolts fully.

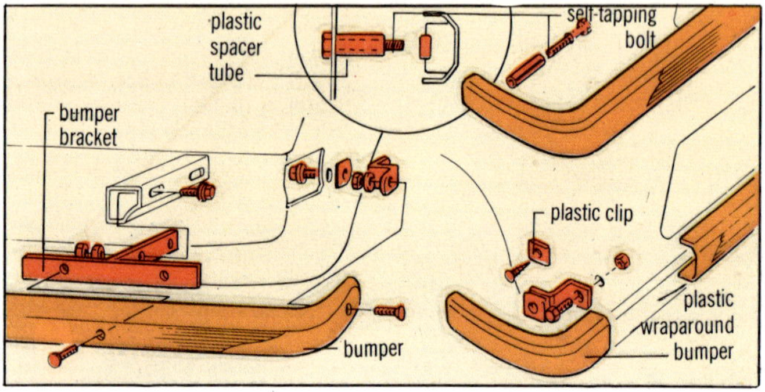

1. Bumpers are fixed to the car in several places

2. Undoing the mounting bolts

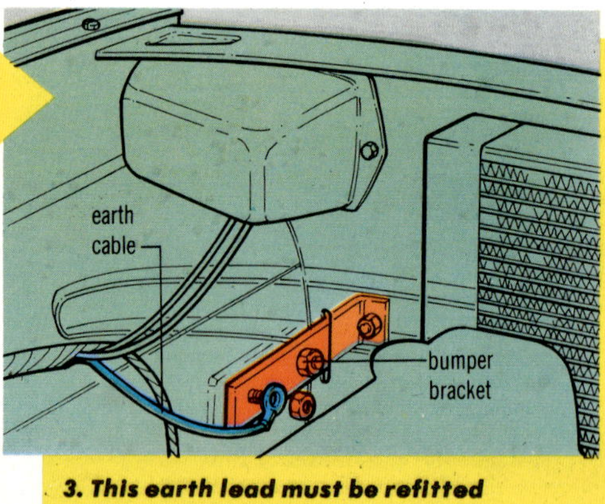

3. This earth lead must be refitted

ALTERNATIVE 2 — FIT 'UNIVERSAL' BRACKETS

The 'universal' clamp-on bracket type is designed to clip over the top or bottom lip of the bumper and is clamped in position with two bolts (**fig 1**). This type is not suitable for all bumpers — but if you sell the car and want to keep the lights you can take the brackets off without leaving ugly holes.

Use a tape measure to help you decide where to position the brackets (see Fact File — Extra lights law) — then bolt them in position. With the brackets in place all you have to do is fit the lamps (see Step 4) and wire them in (Step 6).

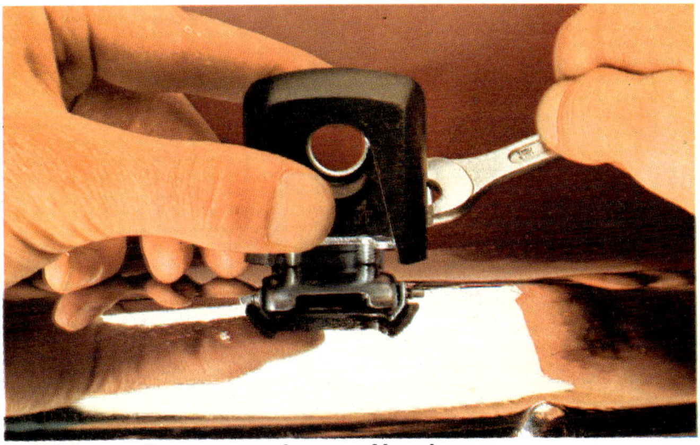

1. Bolting on the 'universal' type of bracket

STEP 3 — DRILL THE BUMPER OR BODY

With the bumper removed, you can now drill the holes for the light mounts. But you must only drill through metal sections: plastic will not be strong enough. Before you drill any car body sections, check behind them as well to make sure you will not damage anything if the drill breaks through suddenly (see Tip — A corking tip).

Use a centre punch to mark the position of the hole and cover this area with masking tape to prevent scratches if the drill bit slips. Then fit the lights, and bolt the bumper into place.

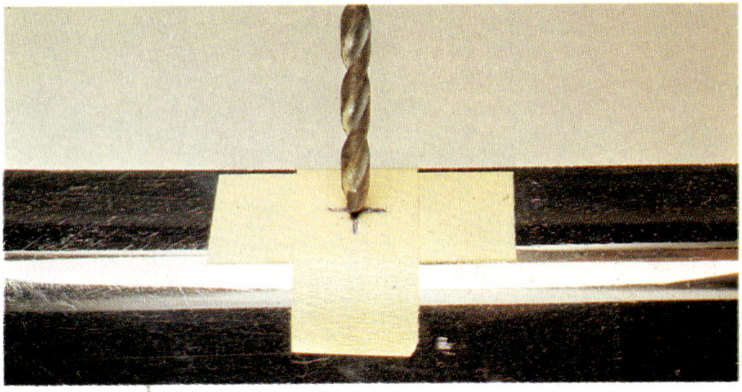

1. Drilling the bumper — be careful not to slip

TIP

Cool tip

When drilling on horizontal surfaces there is an easy way to keep the drill bit fed with oil to keep it cool. Take some Plasticine and mould it into a ring around the point to be drilled. Fill it with oil and drill. The drill bit is automatically kept cool.

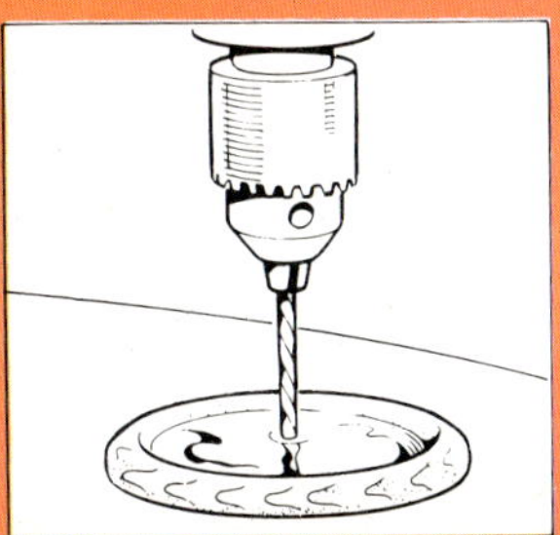

TIP

A corking tip

To avoid the risk of the drill going too far through a panel and damaging something on the other side use this cheap and simple idea.

First decide how far the drill bit can go beyond the panel. Then take one or two corks and force them over the drill bit leaving however much you want of the tip showing. If you have not got any corks your local home-brewing shop will sell you some very cheaply.

Now, when you drill the cork will stop the drill from going too far through the panel, avoiding expensive damage.

STEP 4 FIT THE LIGHTS

All spot and driving lights are fitted with a bolt to fix them in place. Usually this is attached to some kind of movable bracket (**fig 1**) which allows you to adjust the light beam once you have done up the main fixing bolt.

In all cases, the lights simply bolt through the holes in the mounting brackets or the holes you have drilled directly in the bumper or front panel. Alignment is not critical at this stage, but at least make sure that the adjustable brackets are fitted so that they point straight ahead.

You may find that the light mounting bolt sticks down too far and fouls something when it is in place. There are two solutions to this. You can shorten the bolt with a hacksaw, or fit spacing washers (**fig 2**) between the lamp bracket and the bumper or panel. Do not fit too many washers — it can make the lamp unsteady, and allow it to vibrate and upset the light beam. If, when the lamp is fully tightened up, it still seems too sensitive to vibration, you can fit a steady bar (see Step 5).

Protect the threads of the fixing bolts with a coat of zinc primer. Some lights include rubber sleeves which fit over the threads — pack these with rubber grease.

1. This bracket is adjustable

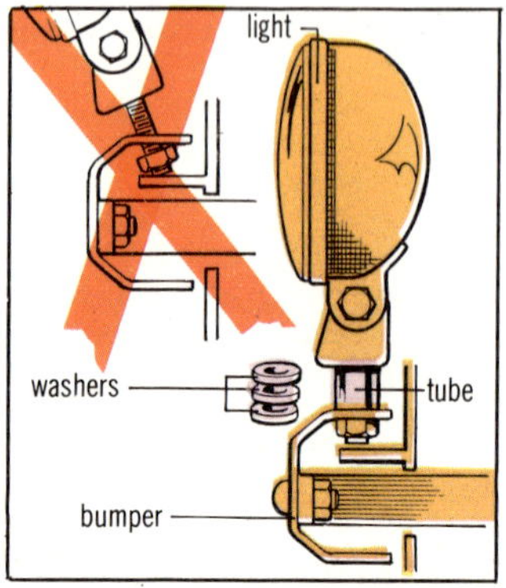

2. Use washers to raise light

STEP 5 FIT A STEADY BAR

To make sure the lights remain steady under all conditions and do not vibrate, you can fit them with steady bars. Some new lights have steady bars included in the kit, and to fit them you should follow the manufacturer's instructions.

However, it is easy to make your own from lengths of ¼ in. (6 mm) threaded rod (available from good ironmongers). You do not have to remove the lights to fit the bars but you must drill a hole in each casing.

First, measure the distance between the light casing and the car body or grille. Cut a piece of rod about 2 in. (50 mm) longer than your measurement. Take off the glass and reflector, then use a 5/16 in. (8 mm) drill bit to drill a hole in the back of the light casing. The position of the hole depends on where you want to fit the other end of the length of studding.

The usual place to fit it is either the grille or the body — if you fix it to the body you must drill another hole for it. Finally, assemble the nuts and washers on the rod as shown in **fig 1** and tighten up fully after the lights have been aligned (see Step 9).

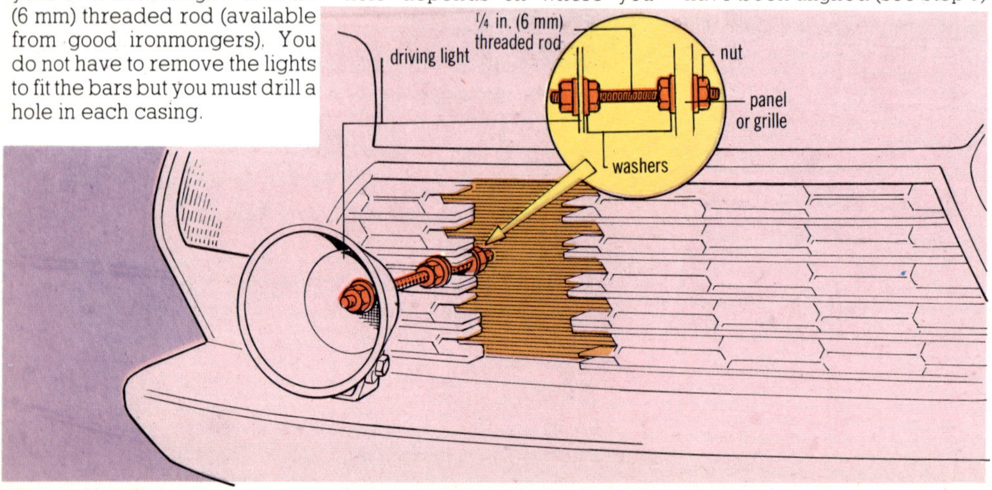

1. How to make up your own steady bar with threaded rod, washers and nuts

STEP 6 EARTH THE LIGHTS

If your lights do not earth through the fixing bolt you will have to fit an earth wire. If you are not sure what type of lights you have take off the front and check the metal casing of the light for a terminal or wire with a spade connector on it. If there is one, the light earths through its fixing bolt. If not, the lights need a separate earth which fits to a terminal near the bulb (**fig 1**).

Take a piece of wire long enough to reach from the lamp to a nearby earth point, preferably somewhere behind the front grille, out of sight. Feed the end of the wire through the front grille then bare about ½ in. (13 mm) of wire and twist the strands together, fold it over and twist again. Now fit a ring type terminal on to it and crimp it into place with pliers or the proper crimping tool — fit the ring terminal under the screw or bolt and tighten it up.

Push the wire through the rubber grommet on the back of the lamp (**fig 2**) and pull it into the lamp. Fit a terminal to suit the earth connection on the lamp — usually a spade connector. Crimp the coloured end of the connector firmly and give the wire a gentle tug to check it is fixed.

Do the same with the other lamp — if the earth point is far from the lamps you may want to fit a single earth for both lamps. Fit the earth wire for one of the lamps as described above, but use heavier gauge cable. Then fit a short length of wire to the other lamp and run it through the grille. Connect the two wires together using a Scotchloc connector — both lamps now use the same earth wire and earthing point.

1. The earth wire fits to this terminal

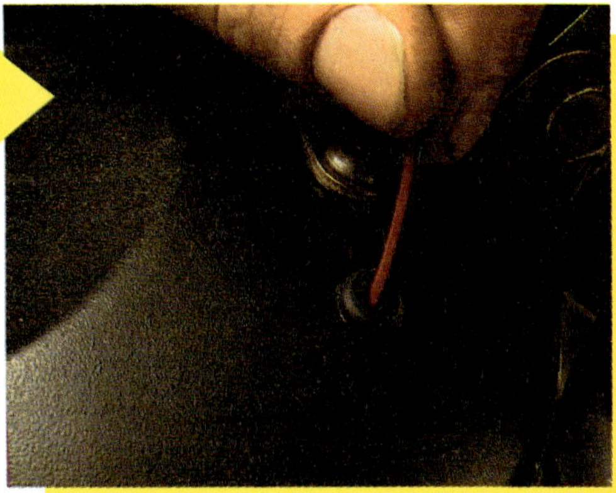

2. Pushing the wire into the light

FACT FILE

Relays

A relay is a special type of remote control switch. It is usually fitted in the circuits of electrical accessories which draw a heavy current. This avoids running thick heavy duty cables from accessories, like lamps or the starter motor, to switches on the dashboard.

The relay is wired into the main power feed for the accessory and carries the heavy current needed while a thin wire goes from the relay to the dashboard switch.

Although relay designs differ they all work in the same way. The main difference is that some have four terminals and need earthing when wired into circuits; others have only three terminals and earth through their metal casing. If you are in any doubt which type to choose follow the manufacturer's instructions for use and suitability.

Whatever type you fit, make sure it is up to the job — the relay rating must be higher than the current drawn by your lamps. Relays are rated in amps, so if you want to fit a pair of spotlights rated at 55 watts (4½ amps) each, then you should buy a relay capable of handling at least ten amps, preferably 12 amps to be really safe.

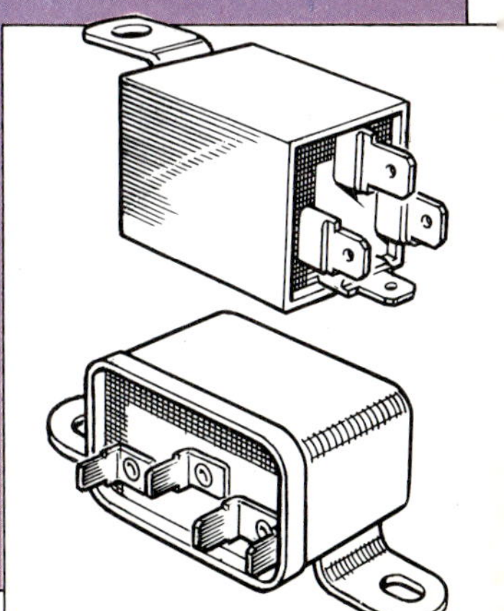

STEP 7 FIT AND WIRE THE RELAY

Before you do any wiring to live feed wires disconnect the earth lead from the battery.

There are several types of relay you can fit — so choose one to suit your lights (see Fact File — Relays) and then find somewhere to fit it in the engine compartment. You should be able to find an existing bolt or screw — alternatively, drill holes and mount it (**fig 1**) with self tapping screws.

Now fit the wires from the relay to the lights (see **fig 3**). Feed the wires either through the grille or holes drilled in the front panel. When you have to drill, check behind the panel first and afterwards paint the bare edges of the hole with primer and topcoat to prevent rusting. Use grommets where the wires pass through the grille or holes in the body panels and make sure the connections are tightly made. Cover any joins with plastic sleeving or tape.

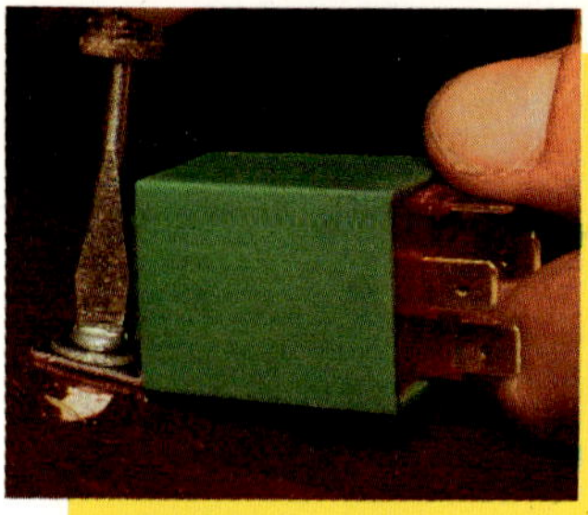

1. Fitting the relay

2. Connecting power lead

Now fit the bulbs. Take great care not to touch the glass as greasy fingermarks may cause bulb failure. If you do touch it, wipe off any marks with a rag moistened with meths.

If you have not fitted a fused relay then the live feed wire has to be fused. Fit an in-line fuseholder (**fig 3**) and secure it in the wire with crimped terminals. A 15 amp fuse should be quite adequate but check to see if the relay manufacturer recommends a certain rating.

On many cars you can take the live feed to the relay from a spare terminal on the starter motor solenoid. Otherwise, connect a large ring terminal to a length of wire long enough to reach the relay and fix it directly to the live terminal of the battery (**fig 2**).

Now move to the other end of the wire, fit an insulating cover and crimp on a spade terminal, and fit it to the relay.

STEP 8 FIT SEPARATE SWITCH

If you want the option of switching the new lights independently of your main beam headlights, you need to fit a switch in the circuit between the main beam power wire and the relay.

First choose a switch to suit your dashboard — simple two-way switches in a variety of styles are available from accessory shops (**fig 1**) — and find a convenient place for it.

If you do not have any blanked off spare switch holes, in your dashboard, either drill a hole in the dash or fit the switch to a small auxiliary panel mounted below the dash (**fig 2**).

Now wire up the switch to the relay and the main beam circuit. Make sure you have enough wire, allowing for slack. The best way to get the wire into the car is to use a single piece of wire, twice as long as a single cable run, folded in two. Then you can force it through grommets more easily.

Find a grommet near to the general area where you have mounted the switch. Push the folded end of the wire through the grommet. Get inside the car and find the cable as it emerges from the bulkhead — use a torch if necessary. Pull the cable carefully into the car while someone pushes it through from the other side.

Cut the loop of wire and bare both the ends. Fit spade connectors and insulators to the wires, then connect one end to the relay, the other to one of the switch terminals.

The terminal on the dashboard switch is wired in to the main beam wire in the same way as described in Step 7.

The lights will now light up when you turn on the headlamp main beam and go out when you go to dip beam. Now align the lights (see Step 9).

1. Switches come in all shapes

2. Fitting switch to dash

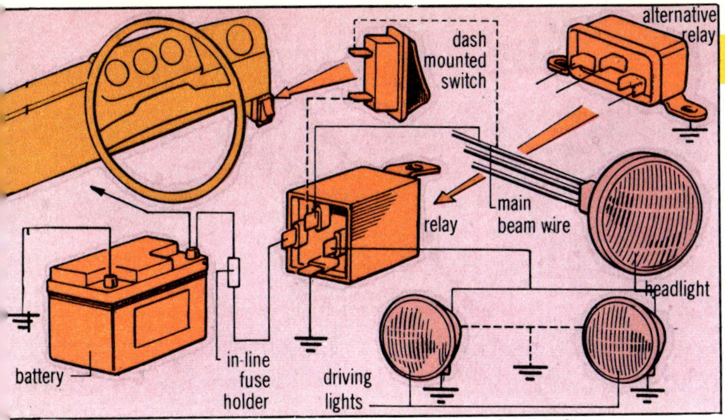

3. How to wire up the lights showing optional switch

You will now be left with one or two terminals on the relay, depending on whether your relay earths through the casing. If it does there will only be one terminal left, but if you have two terminals the relay needs to be earthed. Use a short length of wire and crimp a spade connector to one end and a ring one to the other. Earth the ring connector using the relay mounting screw if possible, and fit the other end to the relay.

You now have to decide whether you want the lights to come on with main beam all the time, or whether you want to be able to switch them off independently. If you choose the latter then you need a dashboard mounted switch (Step 8).

If you choose to have the lights on whenever you use main beam then you only need make one more connection — from the remaining relay terminal to the main beam wire. If you cannot identify the right wire, find the wires as they come out of the back of the headlamp and follow them back into the engine compartment. Now reconnect the battery earth lead.

To find which is the main beam wire, you need a test lamp with a probe. Attach the test lamp's clip to a good earth — bare metal on the car body — then switch on main beam. Probe each of the wires in turn until the lamp lights. Now turn the headlights to dip beam — and if the lamp goes out then you are on the right wire. If not, you have found the sidelight wire — so carry on probing until you find the right one. When you are sure you have the right wire, use a Scotchloc connector to join the relay wire to the main beam wire. Turn on the headlights — the lights will come on as you select main beam. The final step is to align them (Step 9).

STEP 9 ALIGN THE LIGHTS

There are several ways of aligning the lights, depending on what sort of performance you want from them.

If you want them to add extra main beam light then set them so that they illuminate the road just ahead of main beam and overlap into the main beam area.

If you want to light up the outside edges of the road, set up the lights so that they are level with the main beam and point outwards slightly.

Whatever you decide, the adjustment is done in the same way and for best results this should be done when it is dark.

Drive the car about 30 feet from a large door or a wall. Switch on main beam and mark a chalk cross at the centre of the light patch on the wall (**fig 1**). Use a coat or a piece of thick card and masking tape to cover up your headlights so that you are left with only the beams from the new lights.

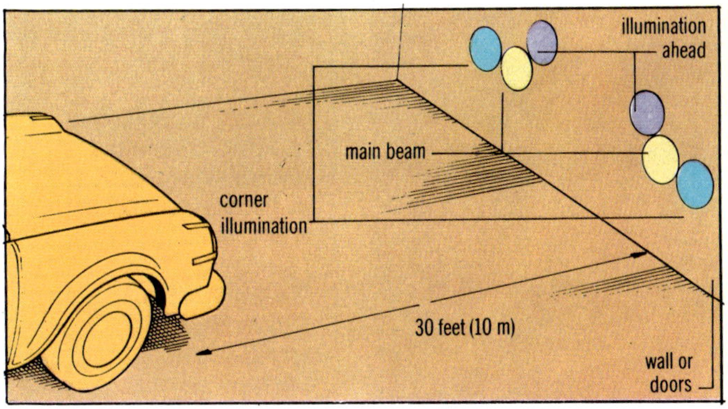

1. Use this diagram as a guide for aligning the lights

See where the beam strikes. Depending on whether you have spot or driving lights, the light will either be concentrated at a point, or spread more widely but still with a distinct bright spot.

Loosen the bolt at the base of the light body and adjust the beam until it is right. If you want the beam ahead of the headlights then aim the beam above the cross — go on adjusting until you are satisfied you have the lamps set correctly and tighten up the bolt. Make sure you do not upset the beam adjustment as you do this. Finally, tighten up the steady bar nuts (see Step 5).

When warning lights fail

When a dashboard light blows it's normally the bulb that's gone. But it could be hiding a more serious fault, so check it now

When a warning light or panel light fails, it is all too easy to put off the job of fixing it until the next weekend — or the weekend after — especially as you may have to remove the dashboard to get to just one small light bulb.

But a faulty warning light may hide the fact that the generator or oil pump has stopped working and you could be left stranded by a breakdown without any warning whatsoever.

In most cases the instruments or dashboard can be removed quite easily to get at the bulb. Remember to disconnect the battery earth lead unless you need power for tests.

When to do this job
When the instrument illumination or dashboard warning lights stop working

What this job involves
Removing the dashboard or instruments
Changing panel light bulbs
Checking for loose or broken wires

Points to watch
Always take more notice of an instrument reading than a warning light
If renewing one of the sender units, make sure that you get the right one or it will never work properly

To do this job
Tools: Spanners; screwdrivers (large and small); small sockets; cranked screwdriver (maybe); torch (maybe); test lamp (maybe); pliers (maybe)
Materials: New bulbs; fine emery paper (maybe)
Time: Varies from car to car; quickest about half an hour; longest not more than four or five hours
Degree of difficulty: Some very easy, others need patience and care when taking out dash

If you have the job professionally done....
Does the light work? Is the dash properly fitted with all the trim and switches in the right place? Do all the instruments work?

STEP 1 IDENTIFY THE FAULT

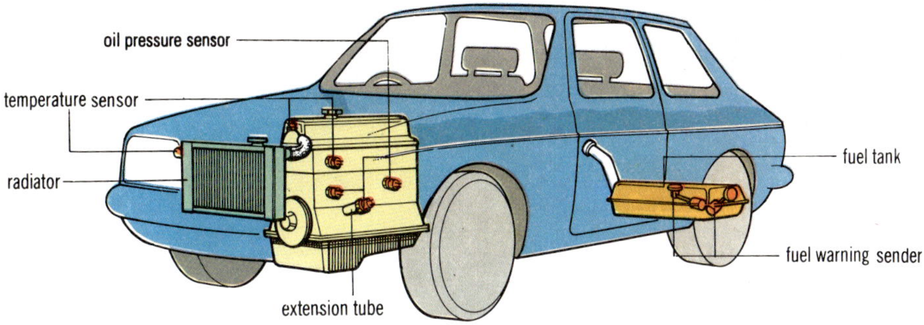

1. Sensors are commonly located in these positions on the engine and fuel tank

If any of the instrument illumination or warning light bulbs in the dashboard fail then there are checks you can make before taking anything to pieces.

If you notice that just one warning light has failed, soon after doing some work under the bonnet, the chances are that you have disturbed a wiring connection somewhere.

The first step is to find the sensor unit (**fig 1**) and check that there are no wires obviously hanging loose — also check that the connector on the end of the wire is firmly fixed.

To test the wiring, take the wire off the sensor and earth it to the engine (**fig 2**). Turn on the ignition — if the lamp now lights the sensor is faulty.

If the lamp does not light and the wiring and connections seem intact, test the wire for breaks between the sensor and the point where the wire disappears into the wiring loom by probing the wire with a test lamp fitted with a pointed end.

Attach the clip of the test lamp to a live point like the battery terminal. Then begin probing the wire at 3 in. (75 mm) intervals.

If the wire is intact then the lamp should light (**fig 3**). If the lamp fails to light at any point on the wire probe back towards the sensor until the lamp lights again — this will show you where the break is in the wire. If the wiring is good then the fault must be a blown light bulb or a wiring fault behind the dash.

If more than one warning light fails then it is probable that the earth is faulty or, if the failed lights all work off one fuse, the fuse may be blown.

If just one of the instrument illumination lights fails then the problem is most likely to be a blown light bulb, but if all fail then it is more likely to be a loose or broken wire, or a faulty light switch or dimmer which is to blame. To cure any of these faults you need to get to the back of the dashboard so see Steps 4, 5 or 6.

TIP

Generator warning light test

If you have a dynamo, take the WL terminal off the control box and earth it. Turn the ignition on — if the light works the control box is faulty.

On an alternator connect a length of wire from the IND multiplug terminal to earth — turn on the ignition and if the bulb lights then the alternator is faulty.

2. Earthing the wire

3. Probing wire for breaks

STEP 2 CHECK YOUR DASH

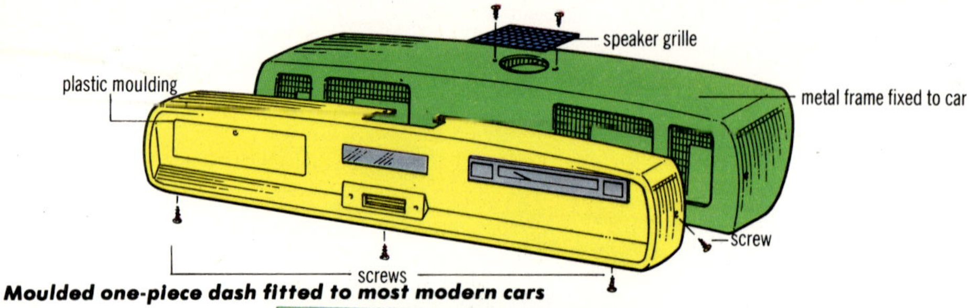

1. Moulded one-piece dash fitted to most modern cars

2. Welded metal type of dash found on smaller cars

3. Once popular, the wooden dash is now fitted only to expensive cars

There are three main types of dashboard and although there are many small differences between cars, the same general method of removal applies to all types.

The plastic moulded type of dash is fitted to most newer cars. These are either of one piece design (**fig 1**) which has to be removed or pod design where the instruments are housed in a pod screwed on to the main part of the dash. The pod type is easier to remove than the one piece dash but the fixing screws are often hidden behind trim so some searching may be needed.

The welded metal type of dash (**fig 2**) is actually a part of the car body. The instruments can usually be removed individually, but some cars have access panels so you can change panel light bulbs without removing the instrument panel.

Wooden dashboards (**fig 3**) are fitted to older or more expensive cars. These can usually be removed in one piece (in some cases the section which holds the instruments comes out on its own) complete with the instruments.

The veneered dashboard looks like a solid wooden one but has a grain effect covering over part of the dash frame.

Before you attempt to take the dash out, see if it is possible to get your hand up behind the instruments from below. Often you can get to them if you loosen or remove the parcel shelf — well worth doing if it saves having to remove the dash.

The parcel shelf is secured at the ends to the side panels of the car. It can be screwed straight on to the panel or bolted on to brackets which are welded to the panel. Some cars have a cardboard finisher screwed under the dash. Undo the screws and you will be able to get behind the instruments and change the bulbs (see Step 7).

If you cannot reach the back of the instruments from underneath then you have to remove the dash (see Step 4, 5 or 6).

STEP 3 REMOVE SWITCHES AND CONTROLS

Some dashboards have switches or heater control knobs which have to be taken off before you can remove them.

Toggle switches often have a chrome plated locking ring with two notches cut in which unscrews to release the switch. Use a small screwdriver in one of the notches and push or tap the ring anti-clockwise. Take great care, the screwdriver can easily slip and scratch the dashboard.

Switches and heater controls have knobs which are held in several ways. The simplest ones pull off but others are held either by a small screw or a spring loaded pip hidden in the side of the knob (**fig 1**). To get these out, use a very small screwdriver to undo the screw or press in the pip.

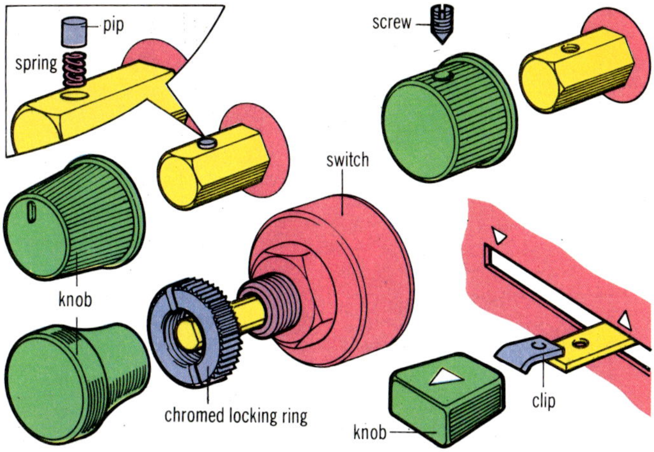

1. Switches and knobs are held in a variety of ways

Once the knob is removed the switch comes off in the same way as a toggle type. Rocker switches can sometimes be removed from the front, but most have clips at the back which have to be squeezed in to release them.

FACT FILE

Dashlight circuit

A typical warning light circuit consists of a sender or sensor unit, which is a type of switch, wiring from the sender to the dashboard and a warning light.

Warning lights which use this type of system include the low oil pressure and low fuel, and high water temperature lights. Alternatively, warning lights can be actuated by a simple switch, like the hand brake warning light.

Warning lights for brake pad wear are different again — the bulb lights when a piece of wire embedded in the brake material is exposed by the pad wearing down. It rubs on the brake disc, earthing the circuit and making the bulb light. Normally these bulbs have a built-in tester activated by pressing a button which earths the circuit and makes the bulb light. If it does not light when you push the button the bulb or the switch is faulty.

Instrument illumination lights are much simpler in operation than warning lights. On most cars they are fed with power from the sidelight circuit — if all the bulbs go out, the sidelight fuse might have blown or there might be a fault in the wiring itself.

A total dashboard light failure is also likely to be due to a power supply or earth failure — possibly a loose connection, broken wire or faulty switch.

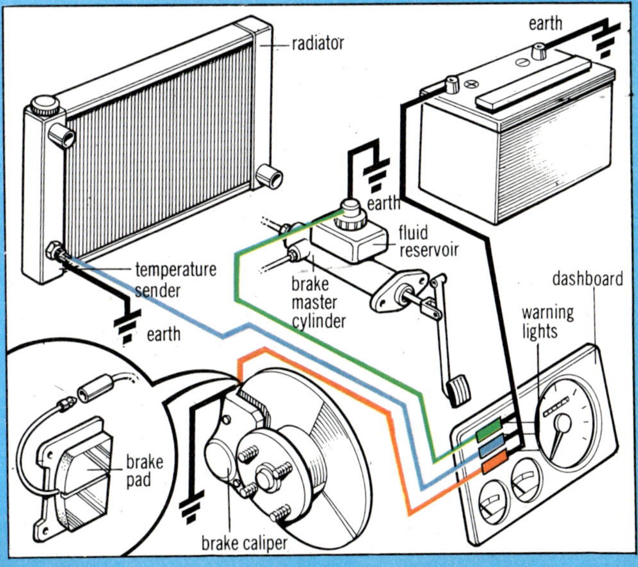

STEP 4 REMOVE A MOULDED OR POD DASH

Nearly all moulded dashboards have the instruments mounted in a separate instrument pod. To get at the instrument or warning lights it is normally possible to detach the pod from the rest of the dash — if not you must remove the dash as a whole.

The instrument pod — sometimes called a binnacle — is normally held by one or more bolts or screws. Sometimes these are easily visible on the front face near the instruments like the Fiat 127 and 128 which have one central screw (**fig 1**). Otherwise they may be under the front lip of the dash or under the hood of the instrument pod as on the Ford Escort (**fig 2**). You may have to hunt for the screws as they are often recessed into holes, or hidden behind small plastic plugs.

In some cases (like the BL Allegro) the instruments do not come out with the pod which is clipped in place and pulls off. Removing the pod simply reveals the instrument fixing screws (**fig 3**). Undo these to get to the light bulbs (see Step 7).

If you cannot find any screws on the instrument pod they may be hidden inside the steering column surround. To remove the surround you will need to find the screws which hold it in place. These are often hidden by plastic plugs which you should prise out — then take out the screws (**fig 4**). The surround will then come apart in two pieces — pull them apart care-

1. The Fiat 127 dash has a central screw fixing

2. The screws in the Escort dash are recessed into the pod

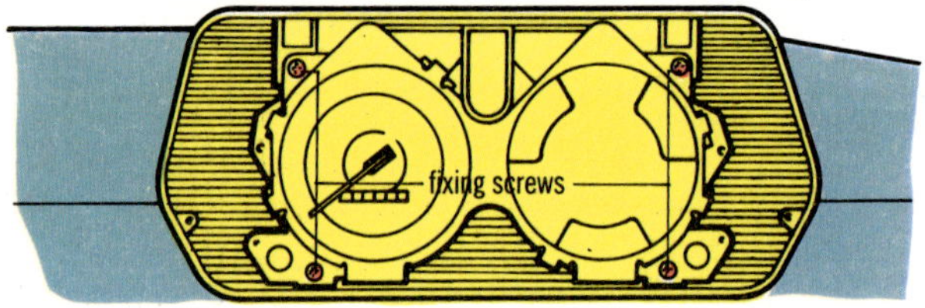

3. The Allegro outer pod unclips to reveal its fixing screws

4. Taking out the column surround screws

TIP

Awkward screw remover

If you have to take out fixing screws which are close to the windscreen or in places under the dash which are difficult to reach with an ordinary screwdriver, try using a cranked or 90° screwdriver. These are available in flat and cross-head types and you can buy them as extensions for a ¼ in. drive socket set.

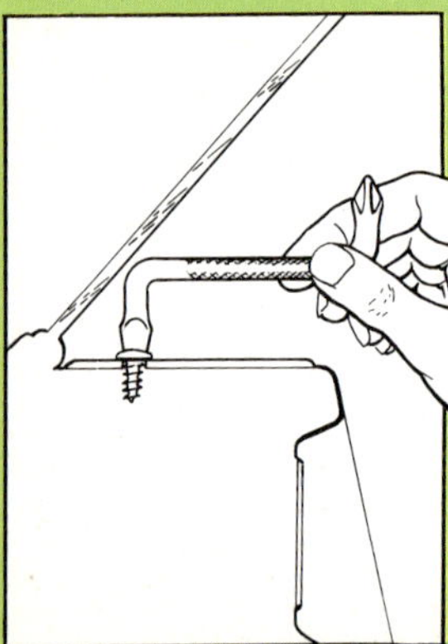

5. **Unscrewing fixing screw from side of dash**

6. **Lifting the dash clear of the frame**

fully as they may have wires attached. Now look for the instrument pod screws which should be visible.

If your car has a moulded one-piece dashboard the procedure for removing it is slightly different. With this type the instruments are fixed into the dash moulding and the bulbs can only be changed once the dash is released from the car.

First make a thorough search for the screws holding the dash. They will be along the bottom edge of the dash and at the ends as well (**fig 5**). Also check inside the glove box if it looks as if it comes off with the rest of the dash. If your dash has a built-in speaker grille on the top face near the windscreen, check to see if it is held by screws. If it is, take the screws out and remove the grille.

Once all the screws are out, try gently tugging the dash forward. Do not use too much force — you may have missed a screw somewhere. If all goes well the dash should pull straight forward (**fig 6**). Be careful not to catch it on the steering wheel or column as you pull it forward.

The instrument wiring should have enough slack in it to allow you to pull the dash forward to change the bulbs. If not, you may have to disconnect the wiring and the speedometer cable. Undo the knurled ring and pull the cable out of the speedometer.

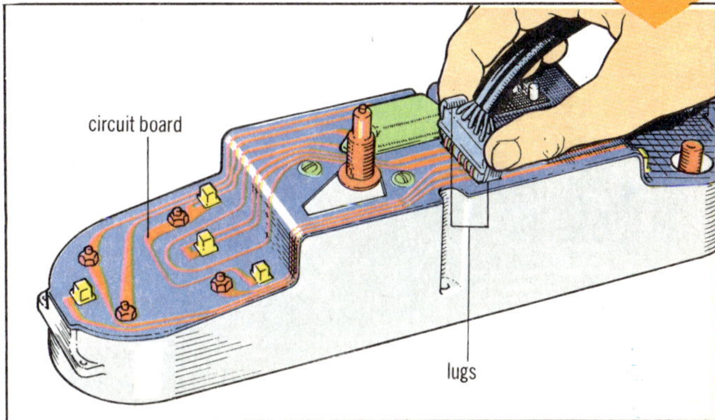

7. **Squeeze in lugs, then pull out the plug**

Each set of wires is held in by a multi-connector plug. These may pull straight out, otherwise look for two plastic lugs which have to be squeezed in (**fig 7**) before you pull the plug out.

Pull the dash forward and you will be able to get at the bulbs (see Step 7).

TIP

One card trick

When you take out the dashboard fixing screws it is all too easy to mix them up. To keep them in the right order take a piece of stiff card and draw an outline of your dash on it. As you take out each screw simply push it through the card in the right position. That way, they all stay in the right place.

STEP 5 WELDED METAL DASH

Older, cheaper cars often have dashboards which are welded to the car body. The instruments come out individually but getting at the instrument illumination or warning lights is made easier on some cars than on others. In most cases you will not have to remove any switches or controls from the dash before you remove the instruments.

The back of the instruments on early Minis which have a central dashboard can be reached through the engine compartment. To get at them you have to remove the air filter box. Pull away the soundproofing and the rubber shroud on the bulkhead. This reveals a hole with a piece of foam rubber in it, which you pull out (**fig 1**). You can now reach the back of the instruments and change the bulbs (see Step 7).

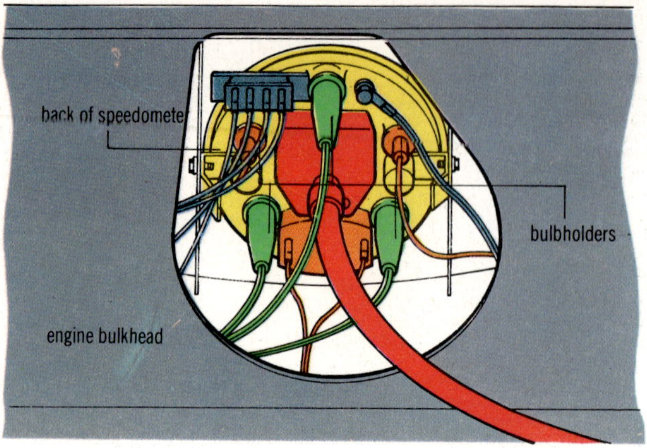

1. Early Minis have an access hatch under the bonnet

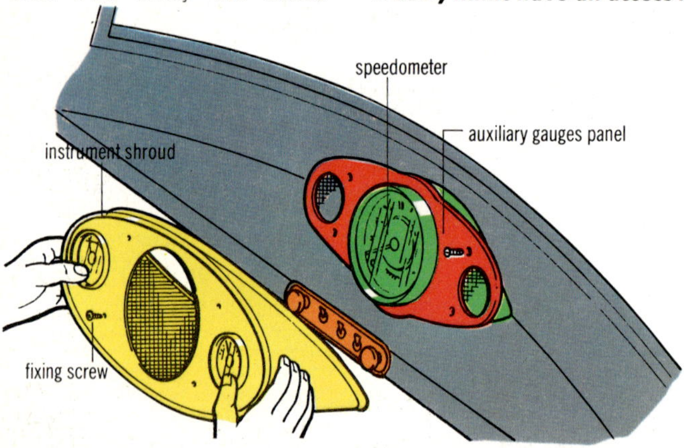

2. Removing the plastic surround from the instrument panel

Later Minis and other cars with a welded dash had better sound insulation — so to get to this type of dash you have to remove the instruments from inside the car. This involves removing the thin plastic trim around the instruments. This is held either with screws or clips (**fig 2**). If there are no obvious signs of any fixings try to prise up one edge of the covering — it is held by hidden clips. Once the covering is off you can see the instrument fixing screws in the panel (**fig 3**). Undo them one at a time and pull the instrument clear — if it does not come out far enough, push the slack in the cable through from inside the engine compartment and take off the cable and the instrument wiring. Alternatively the instruments may be held by a bracket (**fig 4**) in which case you undo the knurled knobs or nuts.

Some cars with metal dashboards have an access hatch in the glove box. This may be a metal plate, which screws to the side of the glove box, a

3. Taking out speedometer fixing screw

rubber bung which just pulls out, or a plastic plug which has to be turned and then removed. With the hatch out of the way you can then usually reach in far enough to pull out the bulbholders. If not, you must remove the instrument from the panel. To do this reach through the hole and find the back of the instrument. It is held by two or three knurled knobs (**fig 4**) or nuts so undo them and take off the fixing bracket(s). Pull the instrument forward from the dash until you can see the bulbholder. Change the bulb as described in Step 7.

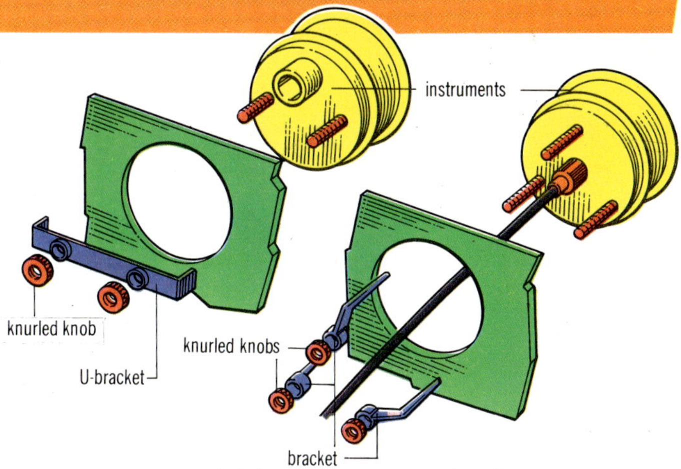

4. Instruments are held in by two types of bracket

STEP 6 REMOVE A WOODEN DASH

Changing instrument or warning light bulbs is made easier on cars with wooden dashboards because you often have enough room behind the dash to change the bulbs without removing it.

If you do not have enough room, you will have to remove the dash or the part housing the instruments, so first check whether you have to release any switches or heater controls from it (see Step 3).

Wooden dashboards are easier to remove than other types because the fixing screws are usually easily visible, either on the front of the dash panel, or at the top or bottom just under the front lip. Take out all the screws one by one, noting the position of any of different lengths — push them through a piece of card to store them safely (see Tip, One card trick).

The whole dash then pulls forward (**fig 1**). Sometimes the top of the dash has pegs which locate into the top finisher. In this case the bottom of the dash tilts forward and pulls down.

Pull the dash forward far enough to get at the instruments. Take care not to strain the wiring as you do this. You should now be able to see the back of the instrument and see where the bulbholder is (for bulb replacement see Step 7).

1. Pulling out the wooden dash from car

If you cannot pull the instrument out far enough to reach the bulbholder, you must disconnect the speedometer cable and the instrument wiring as described in Step 5. On these older cars, the wiring is often connected with plain spade terminals which simply pull off. Make a note of where each wire goes before you take it off or label each one. Once these are undone you can pull the panel forward. To change the bulbs see Step 7.

STEP 7 CHANGE THE LIGHT BULBS

You can now get at the bulbholder. Take it out of the dashboard and replace the bulb.

As **fig 1** shows, there are several different types of bulb and bulbholder which can be fitted to the instruments. Older cars with metal cased instruments have metal bulbholders. These are a push-fit into the instrument — some have spring clips so you may have to pull quite hard to remove them.

More recent cars have instruments with plastic casings, and the bulbholders are either push-fit, bayonet fit (press in and turn) or clip-in. The clip-in sort has two plastic strips on the side of the bulbholder — pinch them together and pull the bulbholder out.

There are three types of light bulb fitted as warning or instrument illumination lights. The bayonet type has a small pip on each side of the cap, which locates into the bulbholder. To remove it simply press the bulb down against the spring pressure, turn it anticlockwise and pull it out.

The screw-fit bulb is removed simply by turning it anticlockwise.

The capless type bulb is even easier to get out. It does not have a metal cap like the other two —

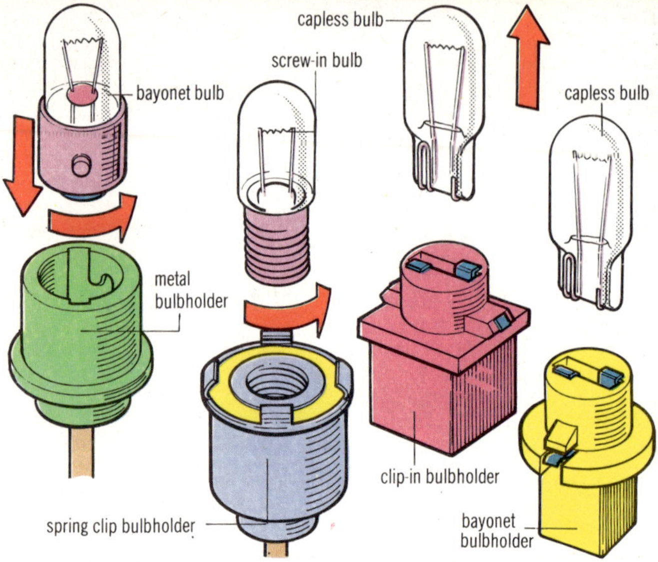

1. Types of bulbholder and bulb — arrows show removal sequence

STEP 8 CHECK THE BULBHOLDER

If the bulb is intact then check the bulbholder.

First check that the bulb is firmly held in the holder. If it is not, bend the terminals closer together so that they grip more.

Also check that the bulbholder is tight in the instrument. On metal-cased instruments squeeze the metal mounting tube (**fig 1**) to give a tighter fit. Bayonet bulbholders should fit tightly in their printed circuit board — if you find a loose one, gently bend the contact strips

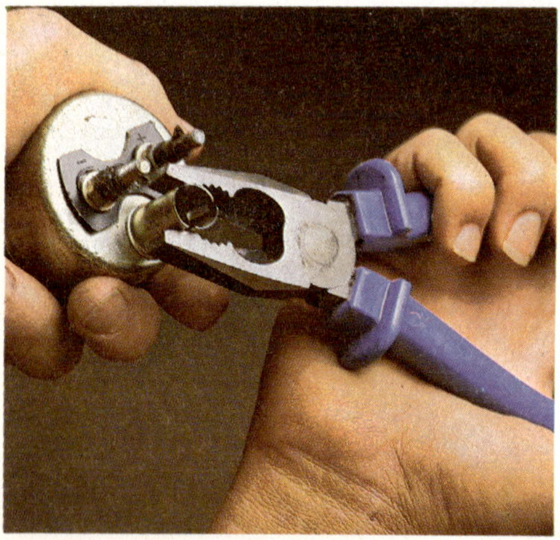

1. Squeezing the bulbholder tube

2. Bending the bulbholder terminals

2. Make sure the terminal strips are in the hole

it simply pulls straight out of the bulbholder. When you replace it make sure that the wires are making contact with the metal strips of the bulbholder — if not, bend them so that the bulb fits snugly.

If the glass part of the bulb breaks off leaving the metal cap in the bulbholder you can remove it by putting a piece of Blu-tack into the cap. Then push the end of a screwdriver into it and pull out the cap.

If that fails use a pair of long-nosed pliers to squeeze the side of the cap in taking care not to damage the bulbholder. Then pull the collapsed cap out.

If the bulb is blown then you have probably found the fault. Before you refit the dashboard test the new bulb by reconnecting the battery earth lead and switching on the light circuit. If it works, refit the dash (see Step 11).

If the old bulb looks intact check it by connecting it directly across the car battery terminals. If it lights, then the fault is in the bulbholder, in the power supply or in the wiring or earthing of the light (see Steps 8, 9 or 10).

When you replace the bulbholder of a printed circuit, make sure that the metal terminals on the holder make contact with the printed circuit on the back of the instruments. Also check that the terminal strips (**fig 2**) of the printed circuit overlap the hole for the bulbholder, otherwise they will not make contact with the holder terminals and the bulb with not light.

(**fig 2**) so that the bulbholder makes good contact.

To check the bulbholder earth and power supply, connect a test lamp between the central connector of the bulbholder and some bare metal that will make a good earth. Turn the power on and if the bulb lights then power is reaching the central connection. This means that the earth is faulty (see Step 9).

If you have the printed circuit type of dash connect the test lamp to the end of the printed circuit (**fig 3**) and a good earth. Once again if the lamp lights then you have power and the fault is in the bulbholder or the new bulb. Check by swapping the bulbholder for one which you know works. If the bulb lights, buy a new bulbholder — if not try another bulb.

If there is no power check the circuit (see Step 10).

3. Checking the printed circuit board for power

STEP 9 CHECK THE EARTH

To test the earth first check that the instrument earth is in place. If the instrument earths through the dash make sure that there is a good contact between the two. If the instrument has a separate earth wire, make sure it is firmly connected at both ends and clean off any corrosion.

Check the earth by clipping the wire on the test lamp to a known live terminal such as the battery — touch the point of the test lamp to the body of the instrument. If it works then the light is earthing properly and the fault is in the power supply to the light (see Step 10).

If the test light does not work then the earth is faulty so make up a new one from a length of wire and connect it between the car body and the instrument casing. The panel light should now work.

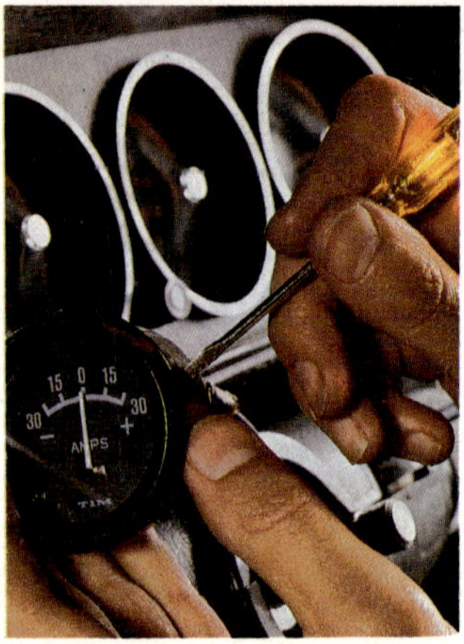

1. Checking instrument earth

If your car has a printed circuit dash, check the earth at the multi-connector plug. Clip the wire of the test lamp to a known live terminal and touch the probe to the earth strip on the plug. If the test lamp lights, the earth is good and the fault is in the live feed to the light (see Step 10). If the test lamp does not light you will have to consult an auto electrician as this type of earth is difficult to fix.

STEP 10 CHECK THE POWER SUPPLY

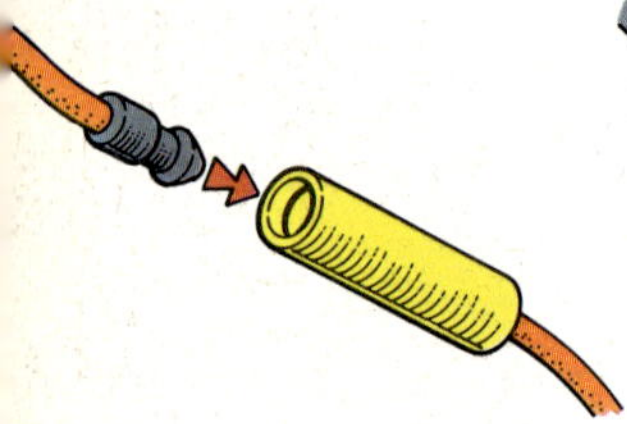

1. Bullet type connectors

2. Multi-connector pulls apart

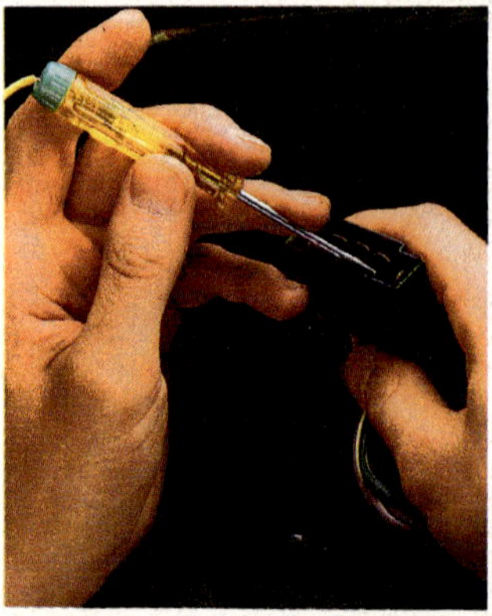

3. Testing the multi-connector

If there is no power at the bulbholder, check the feed wire and its connectors. Follow the wire back from the lamp until you come to the first connector. This may be a bullet connector (**fig 1**) or a multi-connector plug (**fig 2**). Pull the two parts of the connector apart and check the terminals for corrosion. Clean any off with fine emery paper. Reconnect the plug and try it again.

If it fails to work, connect the test lamp between the connector and earth (**fig 3**). If the bulb lights then the wire between the connector and bulbholder needs renewing.

If it does not light then the fault is in the wiring loom and you should consult an auto electrician.

If you have the printed circuit type of instruments, the connector is a multi-connector which is held in place by a slight spring pressure (**fig 4**).

Check in your car handbook to find out which wire connects to the failed light. Clip the wire from the test lamp to a good earth and touch the probe to the correct strip on the connector. If the test lamp lights then the fault lies in the bulbholder (see Step 8) or printed circuit board. If the bulbholder proves to be working then the circuit board is faulty — consult an auto electrician who may be able to fix it.

4. Pulling out the circuit plug

Brake lights not working

Have your brake lights stopped working? If so, a few simple checks with only a minimum of tools should soon pinpoint the fault

Not only is it against the law to drive without your car's brake lights working, but it is also extremely dangerous. So it makes sense to check the brake lights at regular intervals. Luckily, the brake light circuit on a car is one of the most basic of electrical circuits, and so rarely causes many problems. A brake light fault is nearly always caused by a blown bulb, but by following a few simple tests, a more serious problem can be pinpointed and quickly mended.

When to do this job
If one or both brake lights do not work

What this job involves
Removing and checking bulbs and bulbholders
Testing for power in circuit
Checking brake light switch
Renewing the brake light switch

Points to watch
Always carry a spare brake light bulb — they burn out very quickly
Mechanical brake light switches are all similar; you can buy replacements at your accessory shop

To do this job
Tools: Screwdrivers; spanners; test lamp; circuit tester; pliers or crimping tool (maybe); brake bleeding equipment (maybe)
Materials: As required: wet and dry paper; wire; bulb(s); brake light switch; WD-40; brake fluid
Time: Simple faults may take up to an hour to trace and cure. If switch is faulty allow at least two or three hours in total
Degree of difficulty: Most faults are easy to deal with, but the brake light switch is often difficult to get at especially if it is hydraulic

If you have the job professionally done . . .
Do both lights work properly? If a pedal operated switch was replaced, does the pedal return fully?

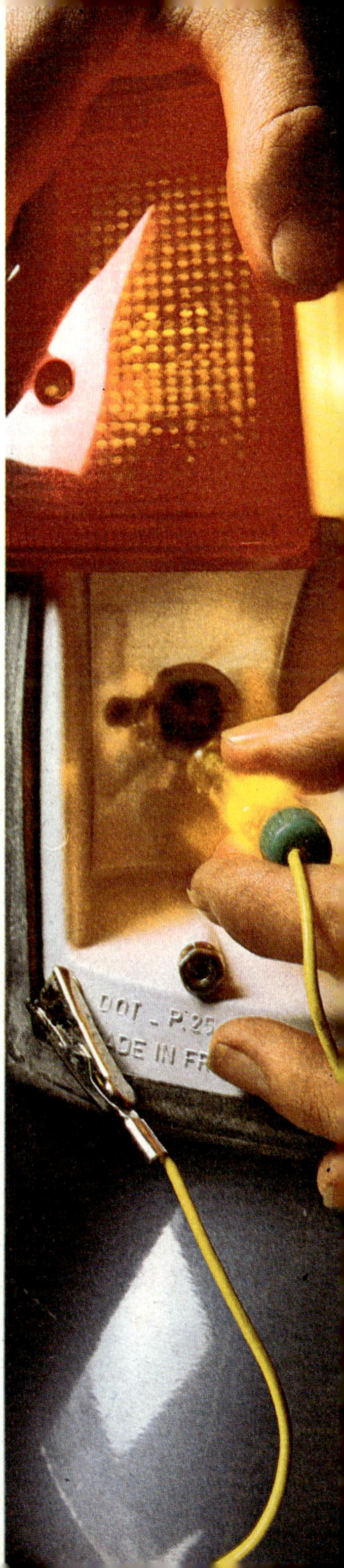

STEP 1 IDENTIFY THE FAULT

When the brake lights give trouble, there are usually clues which indicate where the problem lies. Because they have a very simple circuit, most faults can be traced and put right with a few tools and some simple test equipment. A test lamp does for most jobs though a circuit tester is also handy.

First check the lights. If only one light is out of action the fault is most likely to be a blown bulb as the brake light switch and main fuse are obviously working. However, the fault could also lie in the earth, bulbholder or wire to that light, so if the bulb has not failed check further. Although nearly all cars have only one fuse for both brake lights a few do have a separate fuse for each light. One common fault is that a tail light goes out as you apply the brakes, so no light shows on that side. This is due to a faulty earth, which breaks down when both lights try to earth themselves through it at the same time.

If both lights have failed, then the fault is most likely to be somewhere in the power feed to both lights — the fuse may have blown or the brake light switch may be faulty. It is also possible that both bulbs may have blown at the same time, although this is rare. But if you rarely check your brake lights you may have been running on only one for some time. A simultaneous failure may have been caused by a power surge from the generator. Power surges can occur but if the same fault recurs you should have your generator checked by a specialist.

If the fault is more serious than a blown bulb or fuse and you have no idea of the possible cause, work through each step until you find the problem.

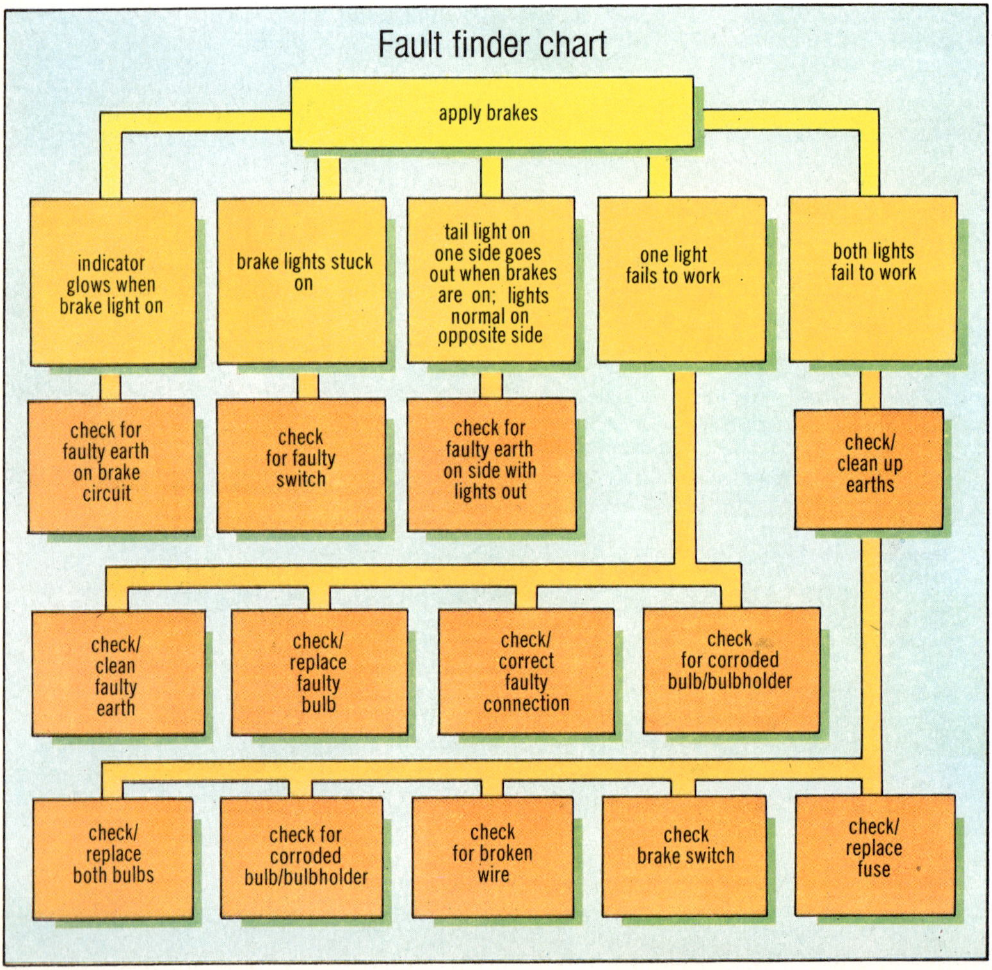

STEP 2 CHECK THE BULBS

To check whether a bulb is blown, you must first remove it from its bulbholder. On some cars the bulb is removed together with its holder from the rear of the light cluster (**fig 1**). In some cases you may have to take the wires off the holder before you remove it, while on others there is enough slack in the wires to get the bulb and its holder out.

On other cars the bulb has to be removed from the front, after the coloured lens has been removed (**fig 1**). Once you have removed the screws, lift the lens away carefully taking care not to damage the sealing rubber. Examine the rubber seal for damage — a split or perished seal will let water into the light unit and possibly result in a short circuit, so buy a new one if it is damaged.

If there are no screws holding the lens, and the bulb cannot be removed from the rear, then the light's rubber housing holds the lens and chromed ring in place.

To remove this sort of light unit, use a small screwdriver to gently prise up the outer lip of the rubber, then work the chromed ring out. If it proves a little stubborn, use another screwdriver to prise the ring free. Now do the same for the lens — be careful not to damage the rubber or it may not make a water tight seal when refitted and the bulbholder will get wet and corrode.

Once the lens is off, you can remove the bulb to check it. Push it inwards then turn it anticlockwise to release it. If the bulb is stuck, spray the holder with penetrating oil and try again. If your fingers slip on the glass, try the Tip — More grip.

If the bulb is stuck very firmly you will probably find that the bulb glass comes away in your hand leaving the metal cap remaining in the bulbholder. If this happens, crush the casing with a pair of pliers and twist the bulb out as usual. If the bulbholder is at all corroded you should clean it thoroughly with wet and dry paper (Step 3) before you fit the new bulb.

If the bulb came out in one piece, check it. Most stop lights are stop-tail lights — the bulb has two filaments, one for the tail light, and the other for the brake. A broken filament should be clearly visible — if you hold the bulb up to the light you should clearly see a break in one of the filaments, but if you are not sure, connect the bulb across the battery using a length of wire. The bulb will have two contacts at its base, one for each circuit; test both of these. If one filament does not light, buy a new bulb: if they both light, the fault lies in the bulbholder or the supply wire.

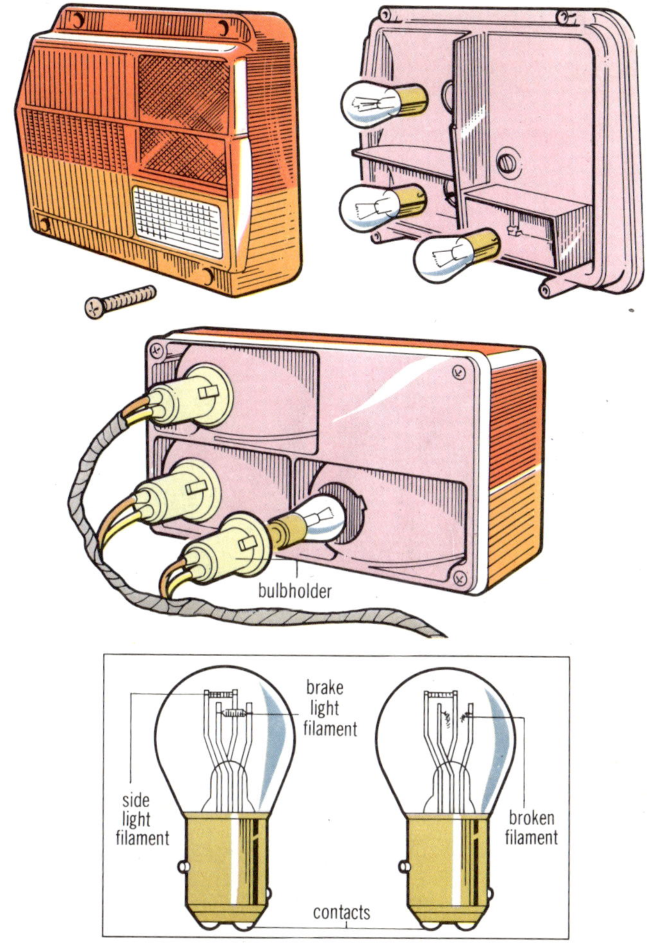

1. Two types of bulbholder (top). Good and bad bulbs (inset)

TIP

More grip

If you have difficulty gripping the bulb then you can get more grip by using rubber gloves. Be careful not to squeeze the glass too hard or it may shatter.

STEP 3 CHECK BULBHOLDER

1. Testing bulbholder

If all the bulbs are intact, then first check in the boot that the multi-plug or any other connection has not come adrift or that any wires have not been torn from their connections.

Now check that power is reaching the bulbholder. There are two terminals in most rear light bulbholders: one for the tail light and one for the brake light. If you are unsure which is

TIP

Wrap it up

If the bulb is not held firmly in the bulbholder, then wrap a little silver paper round the body of the bulb. Make sure that the paper does not touch the contacts at the bottom of the bulb or the bulb will still not light.

2. Cleaning bulb contacts with wire brush

3. Cleaning bulbholder with wet and dry

which, switch on the sidelights and then touch each terminal with the probe of an earthed test lamp. The one which lights is the sidelight terminal. Now touch the other terminal with the test lamp probe and ask an assistant to depress the brake pedal (see Tip). If the test lamp does not light then there is a fault in the power supply wire (see Step 5). If the test lamp lights, then the bulbholder or its earth must be faulty (**fig 1**).

The sort of bulbholder which can be removed from the back of the light can be tested by simply swapping it for the one from the other side. If the light now works try cleaning or tightening the original bulbholder contacts.

With the integral bulbholder, it may just be that there is surface corrosion which is preventing a good contact — clean all traces of corrosion off using an emery board or a piece of wet and dry paper wrapped around a screwdriver. Also check that the bulb is held firmly (see Tip). If you cannot repair the bulbholder, fit a complete new rear light or try to fit just a new bulbholder. If swapping bulbholders does not work, check the earth (Step 4).

TIP

Broom boon

If you do not have a helper handy, the brake can be kept on with the help of a broom. Place the broom handle on the brake pedal, push the pedal down, then jam the brush end in the door pillar corner.

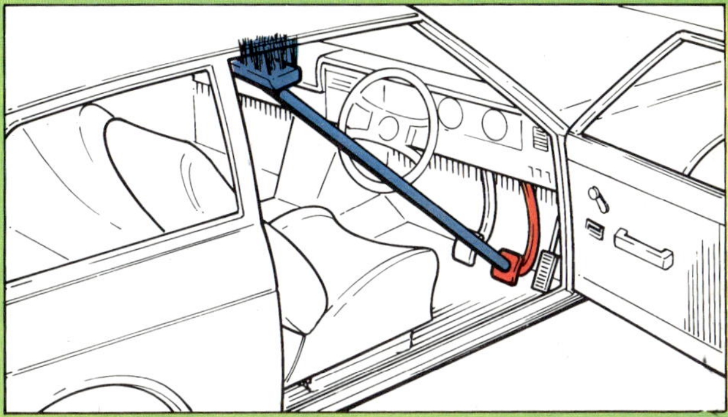

4. Removing corrosion from spade connector

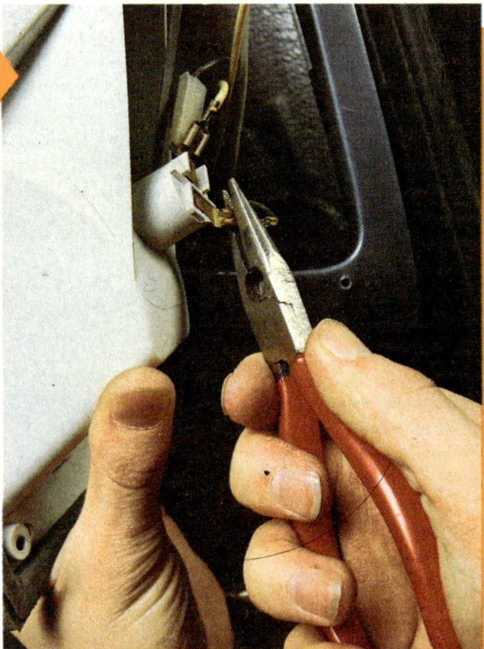

5. Tightening fit of spade connections

STEP 4 CHECK THE EARTH

If power is reaching the central contact in the bulbholder and the bulb and bulbholder are both sound but the light still does not work, then the earth needs checking.

To check the bulbholder's earth, remove the bulb and attach the clip of a test lamp to the earth. This is usually the outer casing of the holder, but on an integral bulbholder there may be a small brass tag which touches the side of the bulb and earths the light. Now turn on the ignition, and press the brake pedal down — probe the brake light terminal in the holder with the test lamp (**fig 2**). If the bulb lights, then the earth is good and the fault must lie in the bulb or bulbholder, so repeat Steps 2 and 3. If the test lamp does not light, then you have confirmed

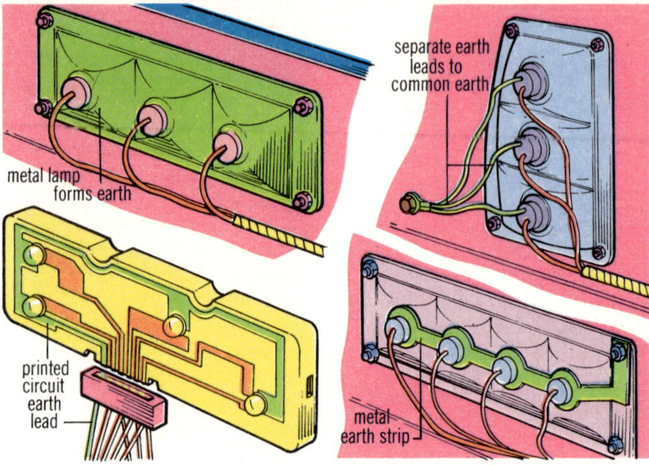

1. Four different methods of earthing the bulbholder

that the earth is faulty.

To repair the earth, you must find out how the bulbholder is earthed to the car body. Metal and clip-in bulbholders usually earth through the casing of the

STEP 5 CHECK THE POWER SUPPLY

If both brake lights have failed and the bulbs are sound or you found no power at the light or at the feed wire in the boot then you must check the power supply from the fusebox to the rest of the circuit.

Your first move is to check the fuse which controls the brake lights (your car's handbook will tell you which one it is). If it has not failed look for any signs of looseness or corrosion. If it appears loose, carefully bend the fusebox terminals so that it is held more tightly (**fig 1**). Corrosion can affect the fuse-box terminals causing either a complete failure of the circuit or — more annoyingly — an intermittent fault. So check the terminals and remove any signs of corrosion with an emery board or wet and dry paper.

If the fuse is securely mounted and not blown, use the test light to check for power at the fuse terminals. The tester should light at both terminals with the fuse in place. If it lights on only one side, recheck the

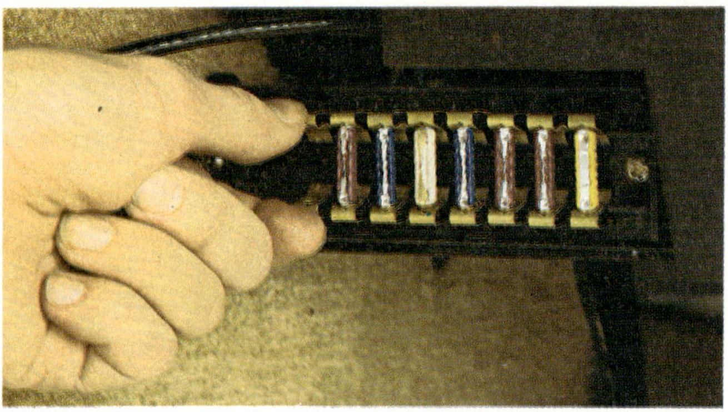

1. Bending fuse terminal for a tighter hold on the fuse

fuse and the holder. If it does not light on either terminal then power is not getting to the fusebox.

You can try replacing the feed wire if the fusebox feed wires are easy to get at. Where the terminals are hidden inside the car the job is a lot more difficult as you will need to get the dashboard out. If you cannot find the wire behind the dash or it is still faulty all the way into the wiring loom, you will have to consult an auto-electrician.

As a temporary measure you can run a suitably thick wire directly from the battery to the fusebox. Be sure to connect it to the right side of the fusebox so that the fuse protects the circuit. If you are in any doubt, once the brake lights are working remove the fuse and check that the lights do not now come on.

If you have power at the fusebox then the next place to check is at the brake light

2. Testing bulbholder earth

3. Cleaning the earth

light. Some other types have a separate wire or metal strip as an earth. Cars with the newer printed circuit type of holder may have the earth built-in to the wiring loom (**fig 1**). If you have this sort then you will have to have the repair done by a specialist.

If your car has the type of light fitting where the light unit bolts to the bodywork or where the earth lead fits to the car body, then check that all the earth points are clean. If not, take out the light unit or release the earth wire and clean up the contact points.

The light unit is normally held to the bodywork by screws, nuts or bolts which can be seen inside the boot or rear wing. Undo these and lift the light unit away. Now clean up all the earth points using wet and dry paper. Coat the earth points with petroleum jelly then refit the light unit or earth lead and check that the brake light now works.

2. Testing for power at the brake light switch

switch. This can be fitted somewhere into the brake hydraulic circuit, where it is operated by pressure in the system, or near to the brake pedal where pressing down on the pedal operates the switch.

To test for power at the switch, attach the test lamp to a good earth point, then press the brake pedal down (**fig 2**). Check for power at both switch terminals — if the test lamp only lights on one terminal, the switch is faulty and must be changed. To confirm this, disconnect the two wires that run into the switch and then connect the two together, thus bypassing the switch. If the brake lights now come on you can be sure that the switch is broken and will have to be changed (see Step 6).

If the test lamp lights when touched to either terminal of the switch then the switch is working and the fault must be in the wire to the back of the car. Instead of trying to find the break in the wire, and splicing in a new section, renew the entire length of wire that runs from the brake switch to the bulbholder. Disconnect the battery, disconnect the old wire, then lay the new wire down alongside it or under the carpet if that is easier.

Wherever you lay the new wire, make sure that it is not exposed; even if it is under the carpet, tape it to the floor. Connect up the new wire using new connectors where necessary, then connect up the battery and check that the brake lights now work.

No power at either terminal on the switch means that the power wire to the switch is faulty and may need to be replaced. As a double check, connect an extension wire from the fusebox directly to the brake light switch. If the lights now work when the pedal is pressed, renew the wire from the fusebox to the switch.

STEP 6 CHANGE PEDAL-OPERATED SWITCH

If your fault-finding showed the brake light switch to be the culprit, then you must fit a new one. The majority of these switches are sealed units and cannot be repaired, so the only solution is to buy a new one or to get one from a scrapped car.

Most pedal-operated switches are attached to a metal bracket fixed to the pedal box. Sometimes the bracket is bolted to the box by a clamping bolt which can be loosened to adjust the position of the switch. On other cars the bracket is fixed, and the switch is screwed in or out to give the correct pedal-to-switch clearance (**fig 1**).

Pull both wires off the switch, marking them so that they can be replaced correctly. Then undo the clamping bolt or locknut, and release the switch from the pedal box.

The hydraulically operated switch is usually mounted somewhere in the engine compartment and looks like a large bolt with a couple of wires going into it. It is normally fitted at the junction of two or more brake pipes on the bulkhead or to the master cylinder itself. Occasionally the switch is screwed into the brake pressure regulator which is bolted to the chassis somewhere in the vicinity of the rear suspension.

As you screw the switch out, it opens the hydraulic system and so brake fluid will escape and air will enter. Minimise this by stretching a thin sheet of polythene over the filler hole in the fluid reservoir and screwing the cap down over the plastic.

The new switch just screws back into place. If you use a new copper washer to seal the thread, there is no chance of a leak. Make sure that the switch is screwed tightly into place – but not too tight to damage the thread (**fig. 3**).

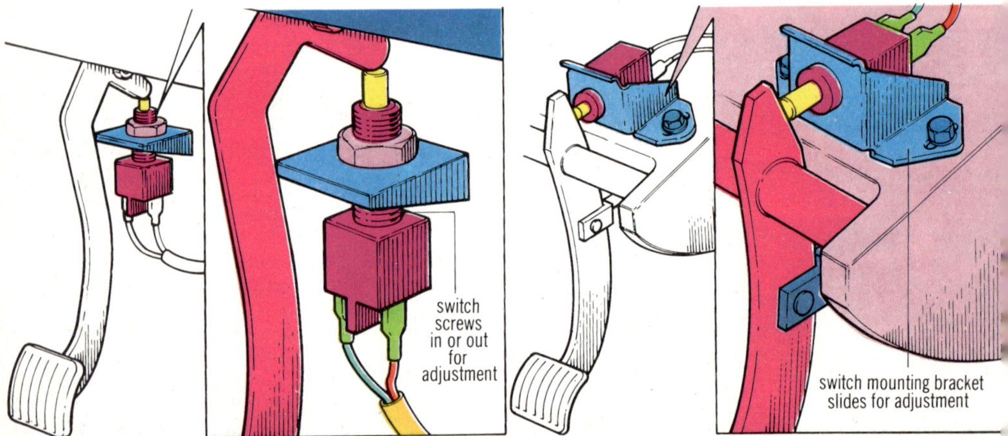

1. Two different mountings for pedal-operated brake light switch — both are adjustable

2. Removing a faulty hydraulic switch

3. Fitting a new switch with new washer

When indicators fail to flash

Have your indicator lights stopped flashing or do they flash at the wrong speed? If so, a few simple checks using the minimum of tools will soon pinpoint the fault

Indicator lights which fail to work properly are annoying as well as being illegal and potentially dangerous.

Solving indicator faults is, however, one of the easiest electrical jobs you can do on the car, so do not be afraid to tackle the repair yourself — even if you have little or no experience of this kind of work — you only need patience.

When to do this job
When the indicators do not flash properly
When the indicators do not work at all

What this job involves
Checking the bulbs
Making sure the lamps are earthing properly
Checking the wiring
Testing the flasher unit
Testing the fuse
Testing the switch

Points to watch
If unscrewing the old flasher unit from the bulkhead proves tricky, just fit the new one to the wiring; tie it up out of the way leaving the old one in place

To do this job
Tools: Flat and crosshead screwdrivers; spanner for lamp unit (maybe); circuit tester; test lamp; sandpaper; thin-nosed pliers
Materials: Plastic electrician's tape (maybe); bulb(s) (maybe); flasher unit (maybe), fuse (maybe); new length of wire (maybe)
Time: 10 minutes to 2 hours
Degree of difficulty: Needs patience and systematic approach

If you have the job professionally done. . .
Do all the lamps now work? Do both sides flash at the same speed?

STEP 1 MAKE A VISUAL CHECK

You can save yourself a lot of time and effort when you are faultfinding simply by making a careful visual check first and identifying the fault accurately.

If only one indicator bulb is not working, you know that the fuse, the switch and the flasher unit are all sound. The fault must, therefore, lie between the switch and bulb and is most likely to be a blown bulb (see Step 3). But you should check all of the wiring (see Steps 6 & 7) and the bulbholder (see Step 5) as well.

If the indicators on only one side of the car work then the fault is most likely to be in the switch (see Step 8) or the wiring (see Step 6) where both bulbs share a single wire at some point. It is much less likely that two bulbs or two separate wires would fail simultaneously.

If none of the indicators are working the most likely cause is a blown fuse (see Step 2), but a faulty flasher (see Step 9) or switch (see Step 8) are other possible causes of the fault.

If another electrical accessory has failed as well as the indicators, suspect the fuse. If the indicators come on but do not flash, suspect the flasher unit (see Step 9). If all the lights work but do not flash, or flash much faster than normal again, suspect the flasher unit or one of the indicator bulbs (see Step 3) — this latter fault is particularly common on cars with additional repeater indicators

STEP 2 CHECK THE FUSE

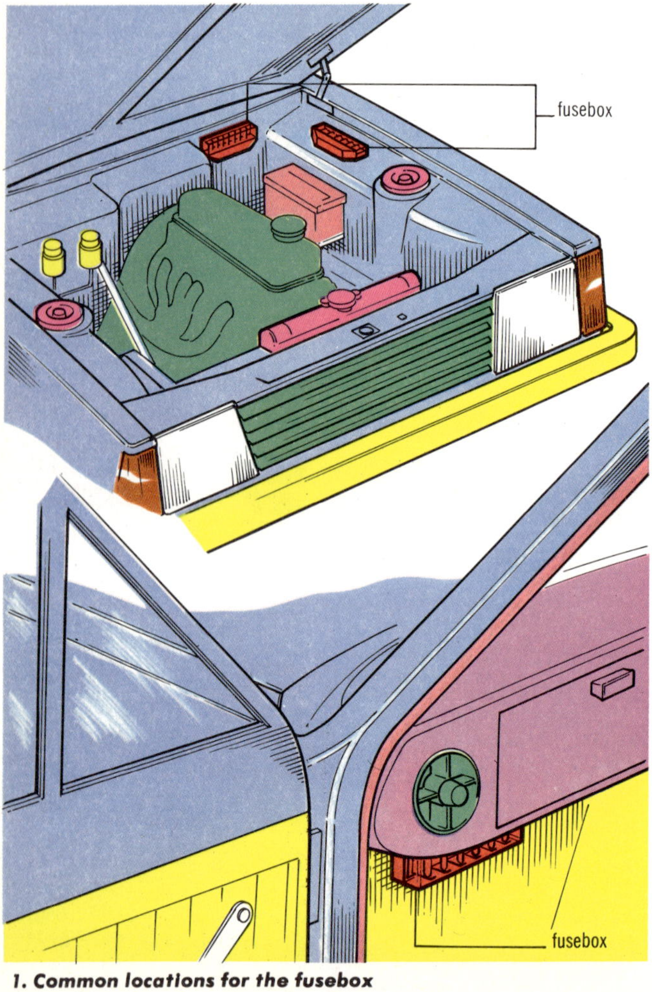

1. Common locations for the fusebox

Your car's handbook will show you the exact location of the fusebox holder. This is a small box usually protected by an easily detachable cover, with the fuses mounted side by side inside. You will usually find it fairly close to the battery, either on the engine or passenger side of the bulkhead, or inside the passenger compartment.

Your manual should also tell you which fuse protects the indicators and which other components are on the same circuit. Often the stoplights, wipers or instruments are on the same fuse — so if any of these are not working as well, the fuse is sure to be at fault.

If there is another fuse in the holder marked with the same rating as the indicator fuse, do a quick test by changing the two fuses over. If the new fuse makes the indicators work again, you will know that either the old fuse needs to be changed, or it was making a bad connection in the holder.

You can test a fuse in its holder with a test lamp. Fasten the test lamp's clip to earth and switch on the ignition. Then touch the test lamp probe in turn to both of the clips which hold the fuse (**fig 2**). If the test lamp lights in both positions the fuse works and is making good contact with the fuseholder. If it lights when touched to one end fitted to the side of the car.

If the indicators on one side work normally while the other ones light up but do not flash suspect a faulty earth connection (see Step 6).

So, exactly how you approach the job will depend on the symptoms of the fault. But if you feel unable to interpret the clues your car is giving, simply work through the steps one by one and you will be sure to identify and fix the fault.

1. If indicator lights stay on suspect the flasher unit

only check that the fuse is making a good contact with the fuseholder.

Alternatively, you can test the fuse out of its holder with a circuit tester. Clip one end of the circuit tester to one end of the fuse and touch the tester's probe to the other end. If the fuse is good the circuit tester will light.

2. Checking the fuse using a test lamp

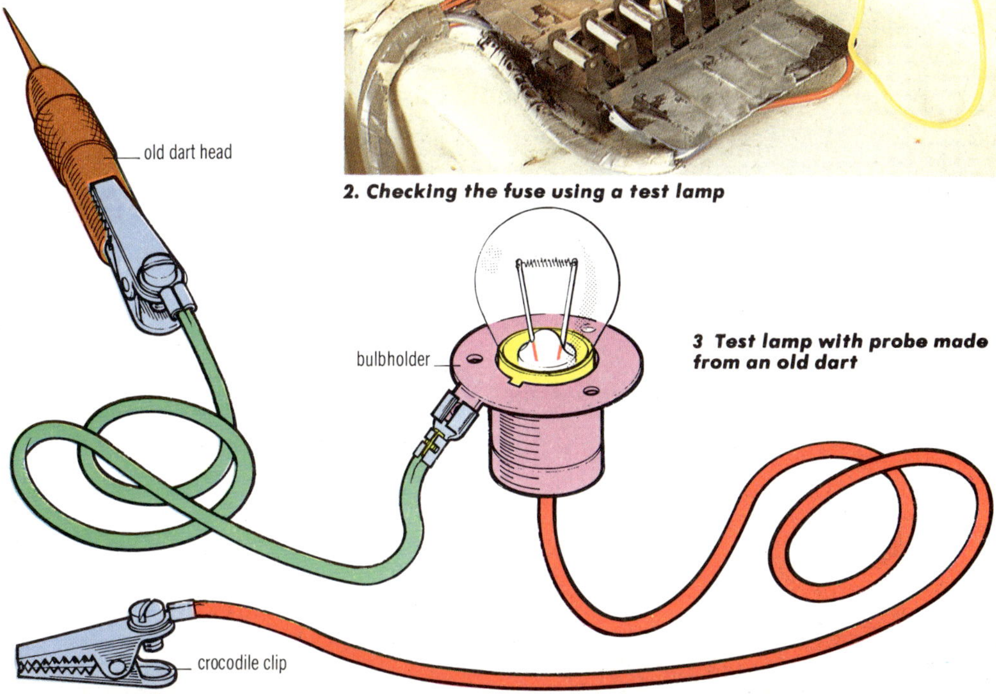

3 Test lamp with probe made from an old dart

STEP 3 CHECK THE BULBS

If one indicator has failed, then a blown bulb is the most likely cause — but even if more than one indicator bulb is not working this quick and simple step is a vital part of your checking procedure.

To check a bulb, first undo the screws holding the plastic lens cover and remove the cover. Push the bulb gently inwards and turn it from the socket. If the bulb is stuck in place, spray the socket with penetrating oil and try to move it from side to side to loosen it. If only the glass comes away when you twist the bulb, you will have to remove the metal part with a pair of long-nosed pliers (**figs 2 & 3**).

You can usually check a bulb visually — hold it against the sky and look for a break in the coiled filament. But if you are unsure whether the bulb has blown or not, test it with the test lamp. Clip the lamp to the live terminal of the battery, press the base terminal of the bulb to the earth terminal of the battery, then touch the probe of the test lamp to the metal body of the bulb. If the bulb does not light, renew it.

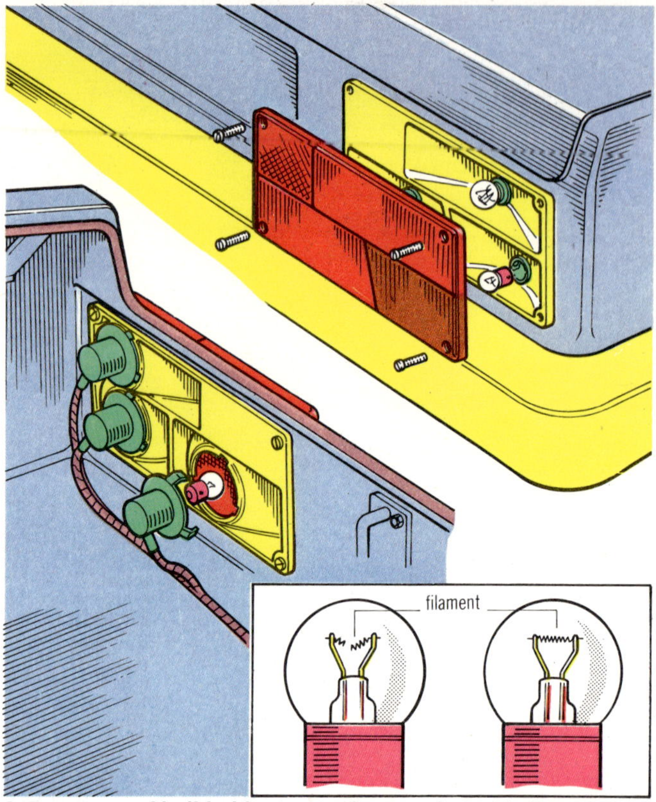

1. Two types of bulbholder (top). Blown and good bulbs (inset)

2. Removing metal part of bulb with pliers

3. Another way of removing a broken bulb

STEP 4 CHECK CONNECTIONS

1. Crimping connector for tighter fit

2. Cleaning up a spade terminal

The indicator may have stopped working because the wires to it have worked loose. To test for this, switch on the indicators and check that the wires to the bulbholder are tight and their connections clean. Do this by pulling the connectors apart and, if necessary, cleaning them with emery paper. If poor contact was the problem then the indicator should start working when you connect the wires back together again.

STEP 5 CHECK BULBHOLDER

Because indicator bulbs are located low down in the front and rear wings they are often exposed to the elements. So even if the bulbs appear to be tight in their holders electrical contact may not be made. To check this, switch the indicators on and try moving the bulb in its holder — removing and replacing it a couple of times may also do the trick. If the bulb flickers or suddenly lights up, the fault lies in the holder.

If the bulb does not light up the fault may still be in the bulbholder. To make sure, you will need to check whether power is reaching it (see Step 7) and whether it is properly earthed (see Step 6). If these connections both prove sound the holder must be at fault. If cleaning it with emery paper fails to have any effect you will have to renew it.

1. Cleaning bulbholder with emery paper

STEP 8 CHECK THE SWITCH

The indicator switch on most cars is located within a cowling on the steering column. This normally splits into two after the removal of a couple of screws. Do not try to force the cowling apart if it appears stuck. It is easy to overlook one of the fixing screws as they are often deeply recessed.

With the cowling apart you should be able to see the switch — it is usually clamped to the steering column or a stalk from it by two screws and a bracket. If you can see the switch terminals without removing the whole assembly, carry out the checks (see below) with it in place.

If you have to remove it, first make a note of the switch position in relation to the steering column, so that you can refit it in the right place (see Tip — On your marks). Now take out the screws and lift out the complete switch assembly. Pay close attention to the three wires for the indicator lights — the input in the centre and the two outputs, one for each side of the car. Make sure they are firmly fixed to their terminals — as they sometimes move with the lever, a wire can occasionally break away from its terminal. In most cases a broken connection can be soldered back into place, but if you cannot do this yourself, disconnect the plug connecting the switch's wires to the loom and take the whole switch unit to an electrical

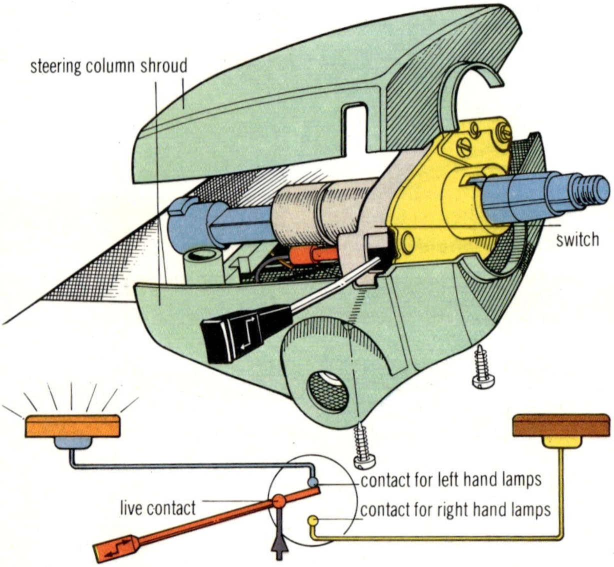

1. Getting to indicator switch (top). How a typical switch works

STEP 9 CHECK FLASHER UNIT

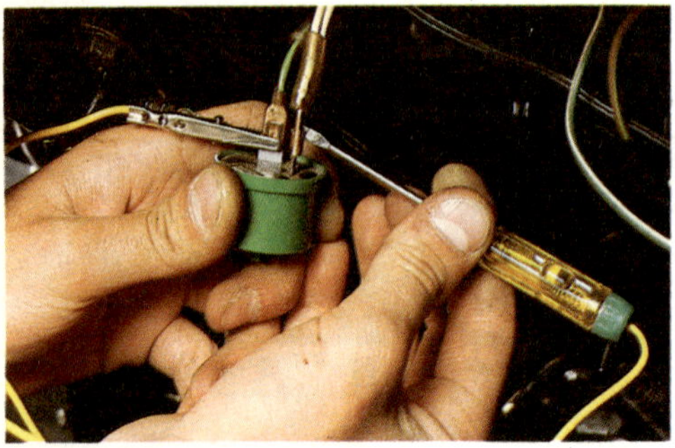
1. Checking the flasher unit

2. Checking power is reaching flasher unit

Your car's manual should tell you where the flasher unit is positioned — but the most common locations are the bulkhead and inner wings in the engine compartment or under the dashboard, screwed to the bulkhead near the steering column.

To test the flasher unit first disconnect the input and output wires and attach them to each end of the test lamp. If the flasher has three leads ignore the central one as this is for the warning light on the dash. Now turn on the ignition and operate the indicator switch. If the faulty indicators now come on it shows that the flasher unit is faulty and should be replaced.

If the indicators do not light check that power is reaching the flasher unit by attaching the test lamp to the power feed wire and a good earth — if the test lamp fails to light there is a power supply wire fault.

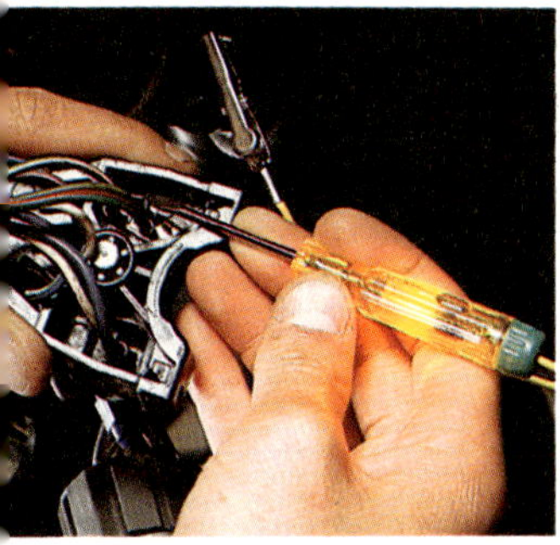

2. Checking switch operation

specialist for resoldering.

Another thing which can go wrong with the switch is that the contacts can become burnt or the moving contact can distort so that it no longer makes a connection. If the contacts are accessible try cleaning the contacts with emery paper and bend the moving contact so that when you operate the switch the moving contact once again reaches the static contact.

To check the electrical operation of the switch, use the test lamp to make sure that there is power going into the centre terminal. Clip the lamp's crocodile clip on an earth point, switch the ignition on and touch the terminal with the test lamp probe. The tester should light if the switch is live. If no current is reaching the switch, check the flasher unit and the supply wire that comes from it.

After you have made sure that the switch is live, turn the switch in both directions and check that power is being switched to each of the output wires serving the indicators. Do this by simply touching the tester's probe to the switch's output terminals (**fig 2**). If the power is not getting across there must be a fault inside the switch. These can rarely be repaired so you will have to fit a replacement switch. New ones are quite expensive so you may find it worthwhile buying a secondhand one from a breaker's yard. Check it before you pay for it as described above but you must use a circuit tester, with its built-in power supply, instead of a test lamp.

TIP

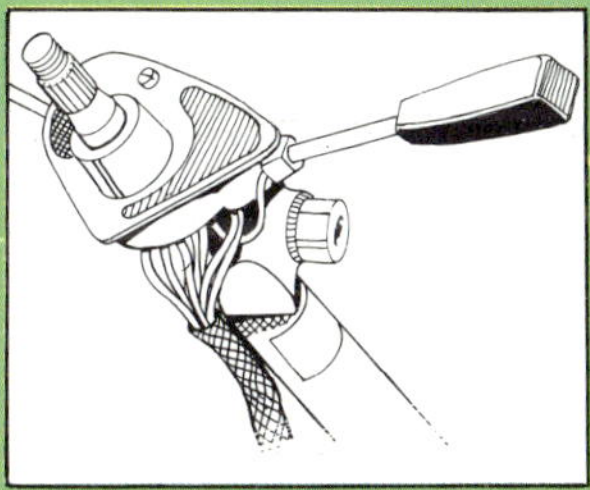

On your marks

Before removing the indicator switch from the steering column mark its position with a felt-tipped pen or paint. This will help you replace the switch correctly so that the indicators will still cancel properly when the wheel self-centres.

STEP 10 CHECK POWER FEED WIRE

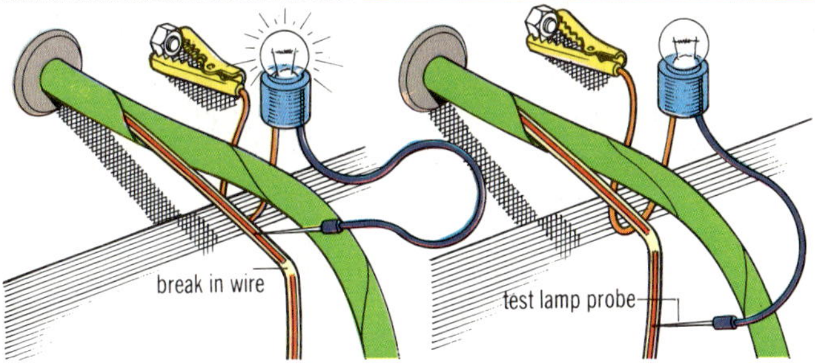

1. Locating a break in the power feed wire

If you know that the switch, flasher unit and fuse are working and live but no current is reaching the bulbholder, the fault must be a break or short-circuit somewhere along the power feed wire to the indicator bulbholder. But the same checking procedure also applies to faultfinding along wires between any two components.

The car's wiring is nearly always colour coded so this will help you to trace wires. Often a wire will disappear through a bulkhead or into a large cluster of wires which are bundled together, but generally a wire stays the same colour along its whole length. A few cars do, however, have numbered wires — each wire having a number at either end. These are harder to trace and may have to be replaced in lengths.

Test any connectors in the line either by peeling back the insulation and using the test lamp probe to the connector terminal, or by pulling the connector apart and touching the probe to the connector terminal on the live side. You can test a wire simply by pushing the point of the probe through the insulation at various points along the wire's length (**fig 1**).

As you work along you should be able to track down the exact location of the fault. If the problem is damaged insulation causing a short circuit, you can patch the wire with insulating tape. If a wire has broken, simply cut out or disconnect the damaged section and renew it.

Repairing pre-engaged starters

When the car fails to start it's often the starter motor that's at fault. Luckily, a few simple repairs are all you need to get going

Pre-engaged starters are robust, heavy-duty components designed to give years of trouble-free service. If they do go wrong, they can cost a considerable amount to renew, so it is well worth overhauling your motor if it shows signs of giving trouble. A little time and trouble may save you the price of a replacement. The parts needed are usually easy to get.

When to do this job
When the starter is sluggish
When the starter operates intermittently
When the starter fails to turn over the engine

What this job involves
Removing the starter motor
Stripping starter and solenoid
Overhauling starter and solenoid

Points to watch
If you have a Bendix-type starter, it is very easy to replace the brushes; re-built Bendix starters are so cheap it is not worth trying to re-build them
New solenoid end caps are often available as part of a re-build kit for pre-engaged starters

To do this job
Tools: Spanners, sockets, feeler gauge (maybe), screwdriver, hammer, pliers, drifts, punch/nail (maybe), small universal puller (maybe), heavy duty soldering iron (maybe)
Materials: Clean rags, methylated spirit, glasspaper, replacement parts: brushes, solenoid, drive gear, bearings (maybe)
Time: Three to four hours
Difficulty of job: Position of starter may make it awkward to remove. Overhauling straightforward, if a little fiddly

If you have the job professionally done...
Does the starter operate efficiently? Does the starter engage immediately you turn the key? Does the starter disengage immediately you release the key?

HOW IT WORKS

Pre-engaged starter

The starter motor is basically a simple yet powerful electric motor mounted on the side of the engine close to the flywheel. At one end of the motor, running from the mainshaft, is a drive gear or pinion. Once the starter motor is switched on, the drive gear engages with a series of teeth running around the outer rim of the flywheel and the motor turns the engine over until it fires.

On the more traditional Bendix starter, the drive gear only engages with the teeth of the flywheel once the motor has started to spin — with the result that the drive gear or flywheel teeth often become damaged.

The pre-engaged starter, on the other hand, only starts to turn once the drive gear is fully engaged. This is done by a solenoid fitted with a plunger mechanism on top of the starter. Attached to the plunger is a fork-shaped lever arm which pushes the drive gear into mesh with the flywheel. Only when the drive gear is fully engaged does the starter motor, activated by the moving plunger, begin to turn.

The pre-engaged starter has a one-way clutch which allows the drive gear to freewheel once the engine has started; this prevents the flywheel, which spins faster than the motor, from turning the motor and damaging it. Once the starter switch is released the spring-loaded plunger pulls the drive gear back out of mesh.

STEP 1 FIND THE STARTER MOTOR

Look for the starter motor low down near the back of the engine, bolted to the bell-housing — that part of the engine which flares out around the flywheel (**fig 1**). It is roughly cylindrical in shape.

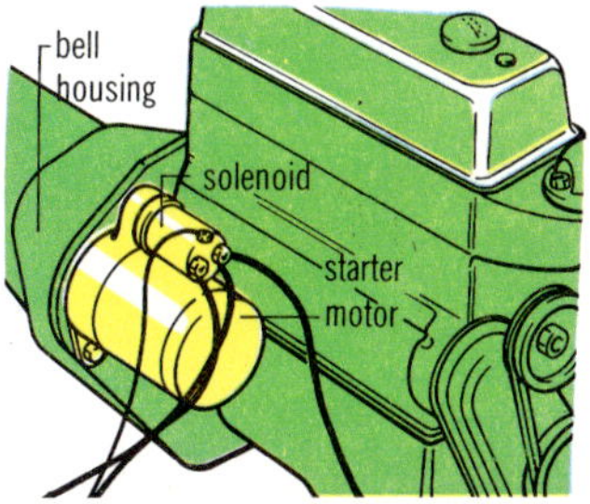

1. Starter motor location

If you cannot find the starter motor immediately, follow the heavy duty feed wire from the battery. This should lead straight down to the solenoid on top of the motor. If the lead goes directly to a separate solenoid — normally mounted on the inner front wing panel — then it is almost certain that the car is fitted with a Bendix starter. This type is not covered here.

If you are unsure which part of the starter motor needs attention, you can narrow down the source of any problem with a test lamp — but disconnect the car's coil first.

Connect one end of the test lamp to the output terminal of the solenoid (the one with a lead to the main starter body) and the other end to a suitable earthing point. Then operate the starter switch with the help of an assistant and see what happens. If the test lamp fails to light the solenoid is at fault and should be renewed (see Step 3). If the lamp lights, but the motor fails to turn over, the fault lies in the drive gear (Step 4) or the motor (see Step 5) or the brushes are worn (see Step 6).

STEP 2 REMOVE THE STARTER

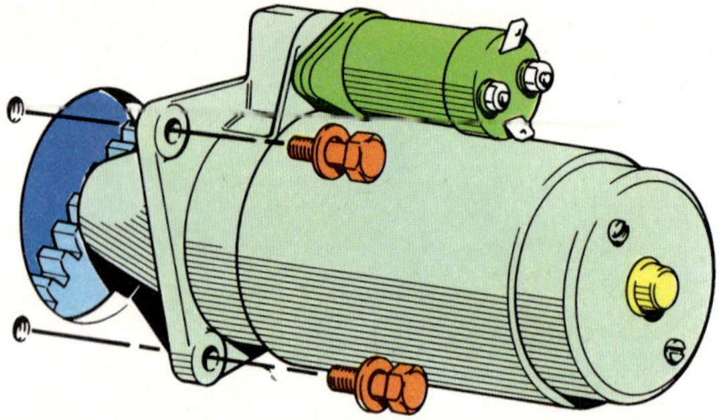

1. The starter is normally secured by two or three bolts

As **fig 1** shows, the starter is normally secured by two or three bolts fitted into threads cut in the engine housing. Depending on the type of car, you may have enough room to lean down into the engine compartment and remove the starter from above. If not, you may be able to get at the mounting bolts by unscrewing the front grille on transverse engined cars. In most cases, however, the starter will be mounted low down in the engine with other components in the way, so that access is only possible from below. If so, support the car safely on axle stands before you start work. Always undo the bottom starter-mounting bolt first as this makes it easier.

First disconnect both battery terminals, starting with the earth lead, and push the leads well away from the terminals to prevent them touching accidentally.

Then disconnect the leads from the starter body. As **fig 2** shows, there are normally three leads — one from the battery, one from the generator (dynamo or alternator) and another from the ignition switch. The battery and generator leads are usually fitted together and retained by a nut; the ignition lead is usually slotted on to a spade terminal.

Next undo the bolts holding the motor (**fig 3**). It may be difficult to get at some of the bolts because they are hidden

STEP 3 RENEW THE STARTER SOLENOID

Occasionally, the solenoid mounted on top of the starter motor goes wrong and has to be renewed. This is simply a case of removing the old solenoid and securing a new one in its place.

To remove the old solenoid, unscrew the electrical connection between the solenoid and the motor. As **fig 1** shows, this is either a U-shaped metal fork or a braided strap with a ring connector at one end sandwiched between two interlocking nuts.

Undo the nuts (**fig 2**). You may have to use an open-ended spanner on the back nut to stop the terminals from turning. If a braided strap is fitted, remove the front nut completely and pull the ring connector free. If the electrical connection is made by a metal fork, just undo the interlocking nuts — the fork is then free and will remain attached to the starter body once the solenoid is removed.

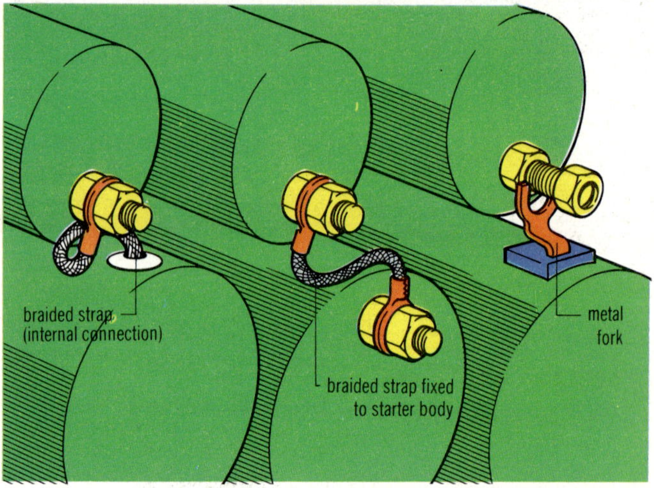

1. Alternative connections between motor and solenoid

Now unbolt the solenoid. This is secured to the motor by two bolts which are fitted into threads cut in the casing. Alternatively, two protruding studs may be fitted to the casing and the solenoid slips over them, and is held by two securing nuts (**fig 3**).

Once the fixings are undone, pull the solenoid free of the starter body. Use a small screwdriver to help you unhook the end of the solenoid plunger

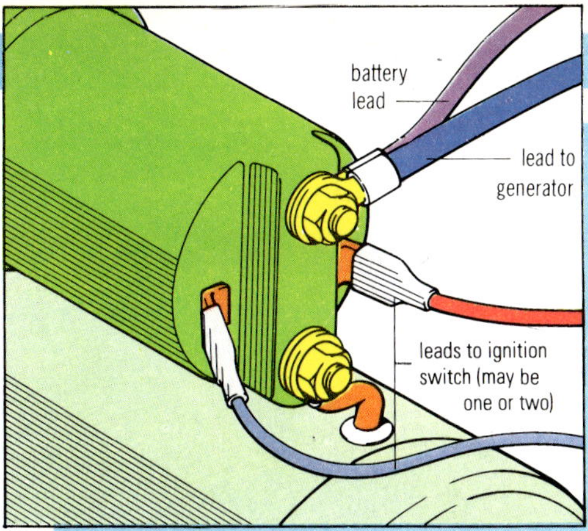

2. Electrical connections to the motor

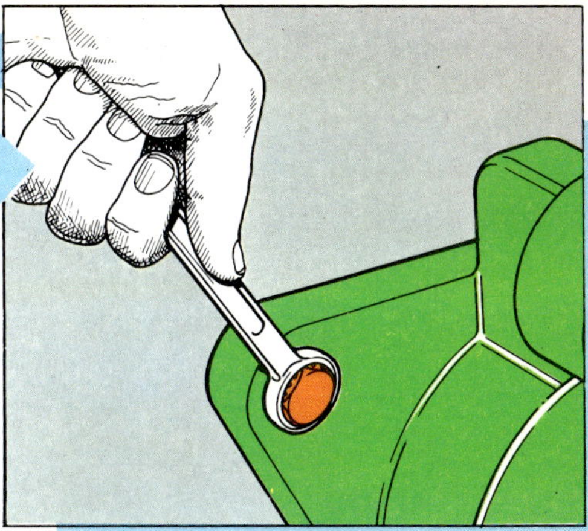
3. Unbolting the motor from the engine

around the far side of the motor or are obstructed by other components. If so, try a socket set with a long extension. Alternatively, it may be possible to work on some of the nuts from above and others from below the car.

The starter motor may drop clear when you release the final bolt, so support it with your free hand. If the motor proves difficult to shift, try twisting the body slowly from side to side pulling backwards at the same time. A really stubborn motor may have to be given a sharp blow with a block of wood to free it.

On most cars there should be enough room to remove the starter quite comfortably, although you may have to twist it slightly in the process. In a few cases you may, however, have to remove one or more other engine components, such as the generator or oil filter, first.

from the top of the connecting arm (**fig 4**).

If you intend to service or examine the other parts of the motor, do this now before refitting the new solenoid.

Fitting the new solenoid is an exact reversal of the removal procedure but, as you push the new solenoid into place, remember to check that the solenoid plunger is correctly engaged with the top of the connecting arm.

TIP

Quick solenoid test

You can check whether the solenoid is working correctly when the starter is removed from the car, by connecting a pair of jump leads to it briefly.

First connect the feed terminal on the battery to the solenoid ignition switch. Then run a lead from the other terminal to the motor body. As the connection is made you should hear the solenoid click into action and see the drive gear move to the engaged position.

2. Undoing electrical lead

3. Unbolting the solenoid

4. Unhooking the plunger

STEP 5 INSPECT THE MOTOR

1. Drum-type commutator (left) and face-type (right)

To inspect the motor, remove and strip it down as described in Steps 2 to 4.

First check and clean the commutator segments (**figs 1 and 2**). If these are badly scored or worn, it is worth buying and fitting a replacement motor as damage or wear cannot easily be rectified. Minor scratches or copper dust can be dealt with without any great problem.

Face-type commutators should be cleaned with a soft cloth moistened with methylated spirit. If you find clinging dirt between the segments clean it out with a sharpened matchstick (see Tip — Picking up dirt).

Drum-type commutators which show signs of light scoring should be rubbed with a piece of fine glasspaper (**fig 2**). Do not use wet and dry paper or emery paper for this job — this will leave a residue which might conduct electricity.

2. Removing light scoring with a piece of glasspaper

3. New commutator (left) compared with old and damaged one

Next check the bearings for wear. Replace the motor casing and endplates. If the endplates wobble on the shaft, the bearings should be renewed as described in Step 7. But check first that the bearing can be purchased separately. If they can only be bought as part of a new endplate assembly then you should think about buying a new or replacement starter motor, instead — it may not work out much more expensive.

Finally, check the brushes. If they are less than ¼ in. (6 mm) long or worn unevenly, renew them as described in Step 6.

While you are examining the brushes, check that all the retaining springs are in good condition and that each brush is held firmly in place. Broken springs should be renewed. If you discover any spring which seems ineffective, it may be possible to bend it slightly until it holds the brush correctly.

TIP

Picking up dirt

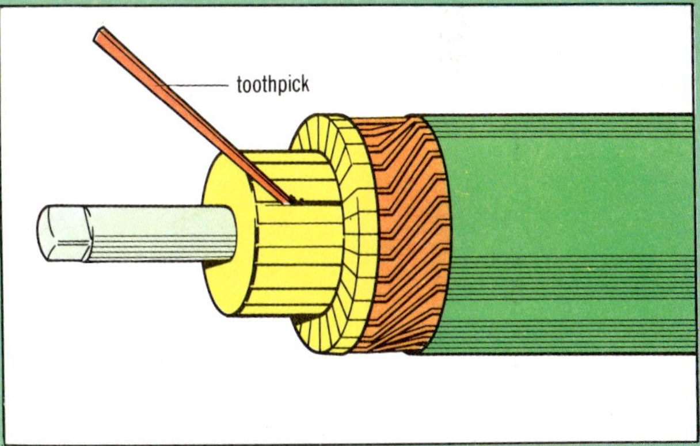

If you find dirt wedged between the segments of a face-type commutator, clean it out with a toothpick or sharpened matchstick. Then brush between the cracks using an old tooth brush. Make sure you get rid of all the loose dirt otherwise the starter may not operate.

STEP 6 RENEW THE BRUSHES

Motor brushes which are worn can be renewed fairly easily. On most pre-engaged starters the brushes are found in the endplate assembly, or in a separate carrier plate underneath. But on some other types you get to the brushes by unbolting or unscrewing a cover plate strapped around the outside of the motor body (**fig 1**). Most starters have four brushes, though some have only two. To make sure you get the correct replacements, it is a good idea to take the endplate or the complete motor along to the shop.

Most brushes are soldered or welded to their feed wires. To refit new brushes to this type you have to cut off the wires of the worn brushes about 3/8 in. (10 mm) from the welds then solder the new brush leads to the exposed wire ends.

Even if you have some experience of soldering this is a fairly difficult job and you may need a heavy-duty soldering iron, so it may be easier to take the parts along to an auto-electrician who will be able to do the job for you.

On other types of starter, the brushes are retained in holders by lightweight springs and the holders are then fixed to the endplate by screws.

To renew these brushes, simply undo the screws, lever back the retaining springs and remove the worn brushes from their holders. Fit new brushes in their place, refit the retaining springs and screw the holders back into position.

Make sure that the brushes slide freely in their housings — rub them with fine glasspaper, to get a good sliding fit.

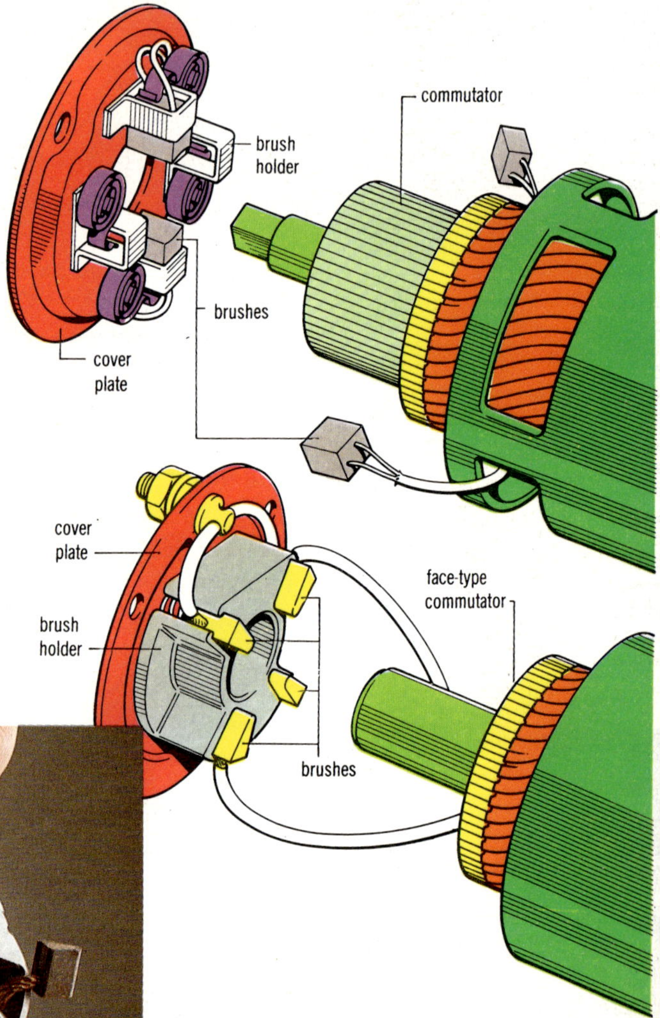

1. Types of brush assembly showing how to remove soldered brushes (inset)

STEP 7 RENEW THE BEARINGS

Pre-engaged starter motor bearings are usually simple bushes made of an alloy of graphite or bronze. Replacement bearings are available for most types of starter but check first just in case.

To remove the old bearing, stand the endplate over a partially opened vice. Use a short length of pipe or a socket spanner as a drift to tap the bearing out of the endplate with a hammer (**fig 1**).

If the bush proves difficult to move, split it down one side with a chisel and then lever it out with a screwdriver (**fig 2**). To avoid damage to your tools use an old wood chisel or a cold chisel for this job.

Soak the new bush in oil for 24 hours before fitting; this will make it easier to press the bush firmly into place (see Tip — Cold treatment). It is best to fit the new bush using a vice, or some other tool which will apply constant pressure. To help the bush slot into place, line it up with a short length of pipe or a socket spanner of the same diameter (**fig 3**). Then tighten the vice until the bush is pressed fully home. Check that the bearing runs smoothly, by fitting the endplate over the armature shaft and giving the plate a few turns.

1. Tapping out a worn bearing with a drift

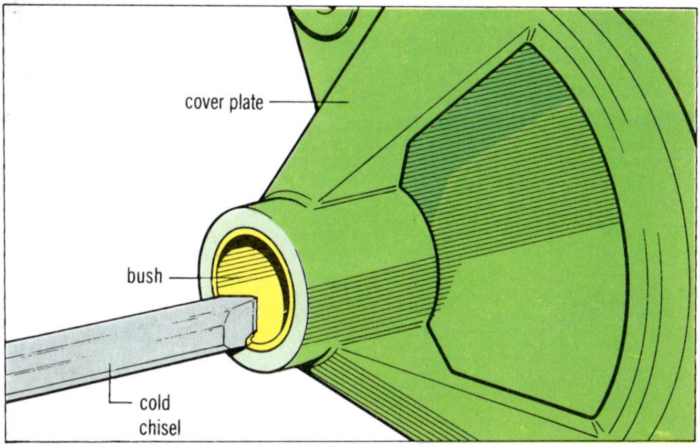

2. Stubborn bearings can be chiselled out

3. Driving the new bearing home with a hammer

TIP

Cold treatment

If the bush proves difficult to fit, check the outside for burrs and rub down with a piece of wet-and-dry paper, if necessary. Bathe the bush in oil and store in the refrigerator or freezer for 24 hours before fitting. It will also help if you carry out the job of fitting the bush in a warm environment so that the endplate has a chance to expand slightly.

STEP 8 REASSEMBLE THE STARTER

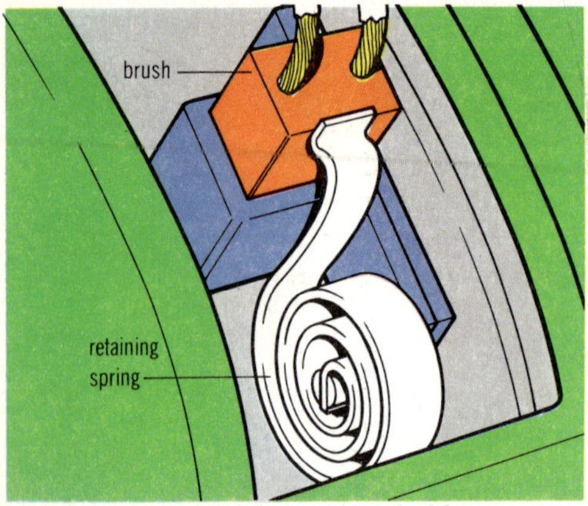

1. Wedging the brushes in their holders

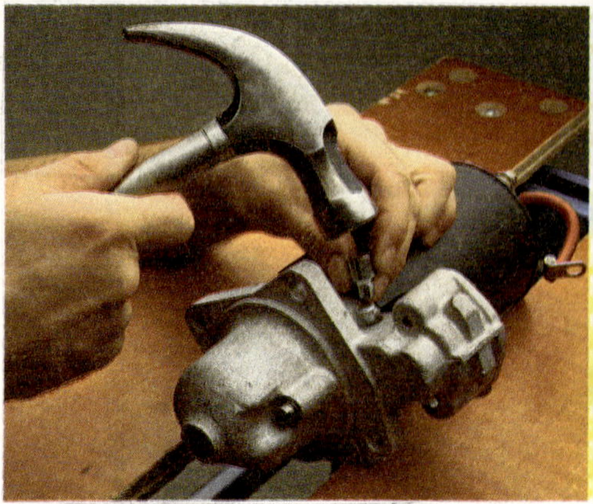

2. Refitting the pivot pin assembly

First fit the motor casing over the armature. On the drum commutator type make sure that all the brushes are wedged at the top of their holders using the retaining springs, otherwise the commutator and endplate will not go back into place correctly (**fig 1**), and you may damage the brushes.

On face-type commutators, simply press the brushes back in the holders with your fingers while you fit the endplate. Then replace the endplate and the drive housing.

To make sure that the endplate is aligned correctly with the motor body there is a cut-out on the edge of the plate which engages with a raised lug on the motor body.

Refit and tighten the through bolts which hold the endplate and drive housing in position. Then screw or bolt the brush inspection cover, if fitted, into place.

Next screw or bolt the pivot pin into place, making sure that it engages with the pivot (**fig 2**).

If the motor is fitted with a screw-type pivot pin, you should adjust the gap between the drive gear and the end bearing. To do this you have to connect the solenoid up to a power source so that the solenoid clicks into its engaged position (see Tip — Quick solenoid test). Insert a feeler gauge between the drive gear and its pinion (**fig 3**). Then adjust the gap by turning the slotted screw. The gap should normally be 0.010 in. (0.25 mm) but check with the manufacturer of the unit if you are unsure.

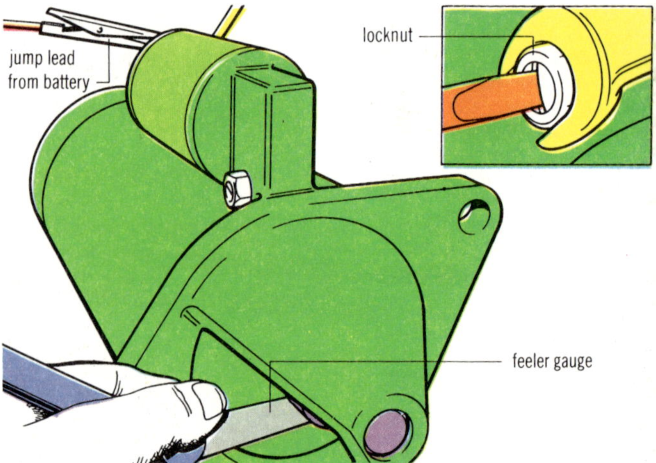

3. Checking and adjusting the screw-type pivot

STEP 9 REFIT THE STARTER MOTOR

Clean the area around the engine where the starter is to be fitted, particularly the aperture where the starter locates. This will help make sure that no loose dirt falls into the engine and that the starter fits squarely against the engine housing.

Offer the starter up, insert the retaining bolts and tighten them progressively in sequence. Before you refit the electrical connections clean all of the terminals and fittings with a rag moistened with meths.

Reconnect the battery terminals starting with the feed wire, then test the motor.

INDEX

Bold entries indicate subjects covered in detail